SUSTAINABLE SUPPLY CHAIN MANAGEMENT IN THE MINING INDUSTRY

SUSTAINABLE SUPPLY CHAIN MANAGEMENT IN THE MINING INDUSTRY

Suchismita Satapathy, PhD
Hullash Chauhan, PhD

AAP APPLE ACADEMIC PRESS

First edition published 2025

Apple Academic Press Inc.
1265 Goldenrod Circle, NE,
Palm Bay, FL 32905 USA

760 Laurentian Drive, Unit 19,
Burlington, ON L7N 0A4, CANADA

CRC Press
2385 NW Executive Center Drive,
Suite 320, Boca Raton FL 33431

4 Park Square, Milton Park,
Abingdon, Oxon, OX14 4RN UK

Library and Archives Canada Cataloguing in Publication

CIP data on file with Canada Library and Archives

Library of Congress Cataloging-in-Publication Data

CIP data on file with US Library of Congress

ISBN: 978-1-77491-584-4 (hbk)
ISBN: 978-1-77491-585-1 (pbk)
ISBN: 978-1-00348-834-7 (ebk)

About the Authors

Suchismita Satapathy, PhD
Associate Professor, School of Mechanical Engineering, KIIT University, Bhubaneswar, Odisha, India

Suchismita Satapathy, PhD, is Associate Professor in the School of Mechanical Engineering, KIIT University, Bhubaneswar, Odisha, India. She has published more than 141 articles in national and international journals and conferences. She has also published many books and e-books for academic and research purposes. Her areas of interest include production operation management, operation research, acoustics, sustainability, and supply chain management. With more than 15 years of teaching and research experience, she has guided many PhD (two completed, four ongoing), MTech (20), and BTech (49) students. Dr Satapathy has published books, such as *Production Operation Management* (Stadium Press) and *MCDM Methods for Waste Management* with CRC Press and *Innovation with IoT and AI* with Springer and many of her projects are ongoing.

Hullash Chauhan, PhD
PhD Research Scholar, KIIT University, Bhubaneshwar, Odisha, India

Hullash Chauhan, PhD, has more than three years of teaching experience at different engineering colleges at both undergraduate and diploma levels. In addition, he has published more than 20 articles in different reputed journals and international conferences. In addition, he has 10 book chapters to his credit. His areas of interest include ergonomics, occupational health and safety, sustainability, supply chain management, and optimization techniques. Dr. Chauhan obtained his PhD in Mechanical Engineering from KIIT, Deemed to be University (Odisha), India. He completed MTech with specialization in Industrial Engineering and Management, and BE in Mechanical Engineering.

Contents

Abbreviations

AHP	analytical hierarchy process
AI	artificial Intelligence
ANFIS	artificial neural network fuzzy inference system
ANFIS	adaptive neural-fuzzy inference system
ANN	artificial neural network
BSC	balanced scorecard
Cd	cadmium
CDP	carbon disclosure project
CSR	corporate social responsibility
CWP	coal worker's pneumoconiosis
DEA	data envelopment analysis
DFS	design for sustainability
DMF	District Mineral Foundation
DT	decision theory
EE	eco-efficiency
ESG	ecological, social, and governance
FA	firefly algorithm
FCM	fuzzy C-mean
FL	fuzzy logic
FMEA	failure mode and effects analysis
GHG	greenhouse gas
GITS	green information technology system
GSCM	green supply chain management
HOQ	house of quality
IBP	incorporated business planning
IoT	Internet of Things
ISM	interpretive structural modeling
KNN	k-nearest neighbors
MAGDM	multiple attributes group decision-making
MCDM	multicriteria decision-making
MSW	municipal solid waste
MT/Year	million-tons/year
NAPM	National Association for Purchasing Management

NDA	negative-distance from average
NMET	National Mineral Exploration Trust
NP	nondeterministic
OC	optimization criteria
OCRA	operational competitiveness rating analyses
OGE	oil and gasoline extraction
OOTGU	optimal performance of thermal generation units
PDA	positive-distance from average
PFSSP	permutation flow shop scheduling problem
PSO	particle swarm optimization
PTS	public transportation system
QA	quality assurance
QFD	quality function deployment
RLMHW	reverse logistics of municipal hazardous waste
SA	simulated annealing
SAW	simple additive weighting
SC	supply chains
SCEM	supply chain environmental management
SCM	supply chain management
SCP	social compliance program
SD	sustainable development
SDF	sustainable development framework
SDG	sustainable development goals
SI_1	social impacts
SSCM	sustainable supply chain management
SWARA	step-wise weight assessment ratio analysis
TFNs	trapezoidal fuzzy numbers
TOPSIS	technique for order preference for similarity to ideal solution
TQM	total quality management
UAV	unmanned aerial vehicles
WASPAS	weighted aggregated sum product assessment
WMA	waste management alternatives
WP	weighted product

Preface

Development in modern times has increased steadily over the past 10 years in many countries, especially large countries such as China and India, impacting most people and fields. Mining operations are booming because of rising mineral demand. Mining companies have expanded their operations and have had fantastic financial results. However, what is considered modern progress in some countries may just lead to an increase in basic production in other countries, with the creation being shipped overseas rather than being used for assembly in the mining nation.

Mining activities can broadly be divided into five major categories: coal mining, metal mineral mining, nonmetallic mineral mining and quarrying, oil and gas extraction, and supporting activities for mining. However, extraction of oil and gas is not viewed as inhumane. Coal, copper, iron metal, and gold are four major mining commodities that provide the greatest revenue. More than 70% of these four minerals are produced in 16 countries.

Despite the debate over hazardous atmospheric deterioration, coal, nevertheless powers the heaters in many projects. Coal accounts for 27% of the world's total energy supply. There has never been more demand for coal, which is a significant mining resource in countries such as China, the United States, India, Australia, Russia, Indonesia, South Africa, Germany, and Poland.

The gauges from the Global Energy Organization (IEA) demonstrate that the usage of coal never stopped increasing; nevertheless, if activities including emotional strategy are considered, this tendency will not continue. Hence, in this book, sustainable supply chain management part of coal mines are discussed with several computational and multicriterion decision-making analysis.

On the basis of prior literature, Chapters 1 and 2 of this book explore sustainability and sustainable supply chain management (SSCM) in the context of transportation and electric generating. Chapter 3 employs Fuzzy AHP and QFD methods to identify major SSCM challenges for mines. Chapter 4 discusses the detrimental effects of mines, along with how sustainability management can be forecasted by taking into account the present supply chain patterns in the mining sector.

The most significant difficulties in the mining sector, such as waste management and supplier selection criteria are covered in Chapters 5 and 6 by taking into account specific analytical indicators to discover the most socially and ecologically responsible solution. Chapter 6 evaluates the risks in the mining business and offers solutions. The final section, the conclusion section offers solutions for present and potential future of SSCM solutions. This book will undoubtedly aid scientists, researchers, and students in their exploration of more technologies that could enhance the planet's social, economic, and environmental security while increasing the productivity.

—Suchismita Satapathy

CHAPTER 1

Introduction

ABSTRACT

Sustainability supply chain management in the mining sector is crucial for ensuring the responsible and ethical sourcing of minerals and metals. Mining operations have a significant impact on the environment, local communities, and workers' health and safety. Therefore, implementing sustainable practices throughout the supply chain is essential to minimize these negative impacts. One aspect of sustainability supply chain management in the mining sector is the responsible sourcing of raw materials. This involves conducting thorough due diligence to ensure that the minerals and metals are extracted in an environmentally and socially responsible manner. It also involves working with suppliers who adhere to strict environmental and labor standards.

1.1 GENERAL BACKGROUND

Sustainable Supply Chain Management (SSCM) in Mining Sector

Mining is defined as the method of extracting raw material from the earth's crust and involves a complete life cycle from scrutiny during production to come closure with provisions for possible uses of land. The expansion of new technology benefits to all main mechanisms of the mining industries: exploration, mining (the physical extractions of materials from the earth), processing, related to physical condition, safety issue, and environmental issue (NAP, 2002). In this framework, the operation of the mining industries is exposing to the effects of various macro ecological factors such as

Sustainable Supply Chain Management in the Mining Industry.
Suchismita Satapathy and Hullash Chauhan (Authors)

globalization process and related to technical, social, financial, political, legal, and ecological changes. One of the main challenges is sustainable development and concepts that, by definition, companies to the needs of environmental protection. The environmental problems of the extractive industries, and particularly the mining industry, are common and malicious. An effective way to address these harmful environmental issues is through organizational practices that involve the broader supply chain. Supply chain practices and their role in the strategy and operation of the mining industry have not been fully addressed. Growing awareness of social and environmental issues has led organizations to adopt sustainable practices. In recent years, organizations have been held accountable for their industrial behavior that influences the environment, culture, and the financial system of their own industry, as well as your supply chain participants. Various stakeholders are putting pressure on organization and their supply chain to take sustainable action to achieve the Sustainable Supply Chain Management goal; however, not all stakeholders have the opportunity to force organization to take sustainability initiative. Several of them help organization to at least become aware of sustainability issues like media/press, while others focus on some level of engagement and implementation of sustainability goals like NGO, consumers, administration, shareholders, society activist, universal competition, etc. Production and consumption in our industrial systems generally begin in the raw materials and mining industries. These activities usually start in emerging countries like Ghana. It is also seen that supply chain in mining operations may have severe ecological and social problems with serious economic consequences. The protection of the environment has been increasingly emphasized by the stakeholders, including social organizations, government, and public. As a result, the establishment of a sustainable supply chain is now part of social corporate responsibility as well as a challenge for companies, including small- and medium-sized enterprises (SMEs). The mineral raw materials obtained are part of almost all of the products we use. The greatest leverage effects in the direction of sustainability are to be expected in the mining country (e.g., a mining company). The consideration of Sustainable Supply Chain Management is very important in developing countries in order to take advantage of the development of the sector. Opportunities and minimize social and ecological issues. Increasing pressure from interest groups, government, and nongovernment agencies is forcing industries to execute Green Supply Chain Management

(GSCM), to implementing for essential growth of the industry, Economic and environmental performance and to ensure business sustainability. Environmentally responsible business practices are receiving rising attention from researchers and practitioners. The number of organizations considering the addition of environmental practices into their strategic plans and operations is steadily increasing. Numerous initiatives have given companies incentives to become more surroundings pleasant. The Concepts related to Supply Chain Environmental Management (SCEM) or Supply Chain Greening are generally understood by the industry as evaluating the environmental performance of suppliers and then only doing business with those who meet the legal standards.

Ecological degradation over the past few decades has considerably increased consumer awareness of environmental issues. As consumers become increasingly critical of the industry's reactive environmental policies, a growing number of companies are developing company-wide ecological programs and "green" (environmentally responsible) products. The number of organizations considering the integration of environmental practices into their strategic plans and day-to-day operations is constantly growing. Many initiatives have provided incentive for organization to become greener. Some of these regulation-driven programs are mandatory, but organizations are increasingly adopting voluntary environmental programs as well. Organizations see many of these environmental programs, which may include technical and governmental development projects, as possible alternatives to gain or maintain a competitive advantage. One area of environmental programming that is gaining increasing importance is that which focuses on the external relations between organizations. To help evaluate the alternatives affecting this relationship, present a strategic decision-making framework that will support management decision-making. Environmentally friendly business practices are under intense scrutiny by researchers and practitioners. Interdisciplinary study has included efforts from management, engineering, physics, and social sciences to examine issues relevant to this subject. Likewise, cross-functional groups within organizations and external stakeholders play a role in decisions related to organizations and the natural surroundings. For this reason, when organizational ecological decisions have to be made, they are necessarily strategic in nature and tend to be more complex. These decisions have internal and external implication for the running of an organization. Green supply chain decisions are one of the most recent issues facing organization with strong internal and external connections. An approach to modeling the dynamic nature of business and its relationship to the natural environment in a decision-making framework is a technique capable of accounting for multi-dimensional strategic and qualitative characteristics. As company moves toward greater global competitiveness, the strategic issues surrounding supply management increasingly demand business attention. Increasing demands for reduced costs, higher quality, better customer service, and supply continuity have significantly increased the reputation of supply management in companies. Supply management is more involved in the development and implementation of strategies and plays a larger role in formulating corporate strategies. International source can be a significant amount of the costs of an organization, in particular for the production of concerns. A method by Monczka and Trent (1995) showed that about 14% of the purchasing volume of US manufacturing companies comes from abroad. These costs appear to be even higher for manufacturing companies in other developed countries. For example, the most recent study by the German sister organization of the Institute for Supply Management (formerly the National Association for Purchasing Management (NAPM)) revealed a share of 23% (BME, 1998). Sustainable development is a model that attempts to shape the communication between the surroundings and society in such a way that advances in prosperity are not accompanied by descent in the environmental and social systems that support life in the future. From this perspective, the Indian mining and mining industries, especially small- and medium-sized enterprises (SMEs), are focused on developing their environmental performance by integrating sustainability practices into traditional supply chain management. However, they struggle with their attempts to analyze and identify Sustainable Supply Chain Management (SSCM) practices to increase sustainable performance. Companies in the modern corporate world cannot overlook environmental challenges. These challenges are now on the agendas of senior management and strategic planning due to increased government regulation and stricter public requirements for environmental stewardship. In order to cut costs and provide better customer service, businesses are integrating their supply chain activities at the same time. These two tendencies are interrelated; businesses must bind suppliers and purchasers to fulfil and even surpass the ecological potential of their constituents and their governments. In the case of mining, the supply chain goes all the way from extract procurement to end-user sale. Supply chain management (SCM) is a strategy for planning, executing, and keeping an eye on daily operations, from the acquisition of raw materials through their organization and accessibility during production to their eventual supply of finished goods. By managing quality products and delivery times, supply chain management (SCM) aims to give the highest degree of customer satisfaction. Suppliers are to consolidate their positions in the mining sector need to know who their customers are and what their needs are. Mining companies are long-time customers and seek long-term relationships with their suppliers. There are numerous ways to approach the supply chain for the mining business. A supplier will decide how best to use the supply chain for that industry based on the products or services they provide. Numerous studies show that local providers do not give the services you want. There is a feeling that local providers are too busy and lack capacity, or there is complacency or monopoly and therefore a lack of service or support. These factors will result in alternate sourcing, which can then lead to international sourcing. Due to their frequent global expansion, mining companies are looking to form partnerships with supplier to meet the requirement of all of their sites. The suppliers who work with the mining businesses now have the chance to understand that when they satisfy and even exceed their local clients' expectations, global prospects will present themselves.

1.2 SUSTAINABILITY

Sustainability' term was first coined within the Brundt land Reported in 1987, with the beginning of ecological safety agencies being set up in the United States. Supply chain management has become a key issue for companies that want to be environmentally sustainable. Supply chain management is a vital part of commerce and is critical to business success and customer satisfaction. These practices help improve financial health, improve customer service, and decrease operational costs. The development of sustainable supply chain management has been the concern of different parties, researcher, and practitioner, as have development in the description of sustainability management in the supply chain. The main basis for the sustainability of the supply chain is inseparable to understand the definition of the supply chain itself. The supply chain is a group of three or more entities (organizations) that are directly involved in the upstream flow of products, services, finance, or information from the source to the consumer. The traditional supply chain supports linkages that serve as a means of converting input into output for an organization. Previous supply chain scenarios faced obstacles related to delivering goods on time in a profitable manner. But the exponential increase in urbanization and globalization changed the way people viewed the traditional approach to work. Currently, the dynamics of the supply chain encourage the adoption of ecological practices and the effective use of human, material, and economic resources, the essence of which is becoming summarized under the term sustainability. In the recent period, supply chain management is a suitable hot topic, opening up new places to profitably meet supply and demand patterns. Instead of the same thing, researchers are also focused on this area to explore new horizons, which improve the powers of the supply chain. To thrive in the competitive environment and maintain significant market share, the supply chain must be reviewed to implement meritocratic business practices. Today's environmental problems become one of the discussions of the global industry. This is reflected with the level of public awareness of environmental products with the level of public consciousness for environmental products and encouraged the industries to apply the concept of environmental awareness in their companies. Components, processes, and information flow required building a supply chain management system at the same time. As a result, there is a government rule in place to raise environmental awareness. So, if a company wants to survive on the global market, it can no longer ignore environmental issues. Combination with supply chain management (SCM) property progressively emphasized in the majority industry through the collaboration of all parties, as well as the government, supplier, and consumers. Response to rapid change resulting from ecological uncertainties and strong demand from government, community, and markets to execute sustainable principles as well as technological development and increasing global competition prompted many companies to execute the plan of sustainable supply chain management. Sustainable supply chain management (SSCM) takes ecological aspects into account in all activities, from raw materials to the final disposal of goods, but not only ecological aspects in the supply chain decision-making process, but also the aspect of financial gain. Focusing on supply chain sustainability entails thoroughly considering and developing the issue from beginning to end, Sustainability must be considered from the time raw materials are produced until they reach the final consumer. The topic of sustainability must also integrate topics and current that goes beyond the core of supply chain management. Sustainable supply chain management (SCM) is the integration of sustainable development and supply chain management, encompassing three dimensions and integrating environmental, social, and economic aspects that give impact to corporate strategy. The model of sustainable development combines the plan of sustainability with the model of development and was first defined in the Brundtland report entitled Our Common Future, which was later adopted by the World Commission on Economic Development United Nations (WCED). It is outlined as the development that meets the requirements of the current while not compromising the flexibility of future generations to fulfil their own needs (WCED, 1987). This definition is the idea for the present still as future analysis within the field of sustainability. The concept of sustainable development is now a widely used concept in many areas of activity affecting human life. It arose in the second half of the 20th century as a response to the dynamic financial growth observed in many country of the world, which was frequently observed in the situation of overly intensive and uncontrolled use of natural resources. Therefore, a study by the UN World Commission on Environment and Development titled Our Common Future was issued in the late 1960s in an effort to lessen this undesirable phenomenon for the world. It ought to be stressed that in the European nation the principle of sustainable development gains constitutional standing and its description will be found within the Environmental Protection Act.

The thought of sustainability was Triple Bottom Layer consisting of social, environmental, and economic layers.

Social responsibility: Taking into account the nature of the mining environment, it primarily means ensuring safe working conditions, but it also relates to the social aspects of mining, including miners' families, the mining environment, etc. the foundation for the long-term development of mining and the utilization of natural resources in the form of minerals. Each of them is distinguished by a diverse and growing set of issues; the solutions bring these vital industries closer to achieving full sustainability. It is a challenge that requires multi-directional support from academia and the mining industry. The basic tools for such support currently include method and technology urbanized in the field of earth sciences and mining sciences.

Economic growth: Economic growth means achieving long-term sustainability in projected production levels and meeting customer needs, as well as achieving the economic efficiencies that result from selling the ore mined.

Environmental: Environment protection of natural resources and the environment means caring for the bed and protecting its resources for its rational acquisition, characterized by savings in its exhaustion. This also means taking action that minimizes the negative effects of the various processes with regard to the extraction of mineral resources in the various forms of geological environment and natural environment on the surface.

Environmental protection is increasingly being emphasized by stakeholders including social organizations, government, and the public. As a result, building a sustainable supply chain has become part of corporate social responsibility and a challenge for companies, including small- and medium-sized companies (SMEs). Given the growing importance of sustainable development in developing countries, the trend of shifting polluting industries from developed to developing countries is concerning. As the world's largest developing country, China has suffered from severe industrial pollution, the main culprits of which are SMEs. Pollution (i.e., the amount of pollutants present in an environment) accounts for about 50% of industrial pollution and shows an increasing trend. Although the Chinese government has enacted laws and regulations to control emissions of industrial waste and toxic gases, these alone are not sufficient to basically solve the trouble. The goal of greener development requires companies, especially SMEs, to reform their business philosophy and invest more in long-term sustainable development. In the event that the domestic market in China gradually matures, the competition between companies has shifted from the whole company to the whole supply chain. Therefore, building a sustainable supply chain and integrating a new management method into daily operations have become top priorities for businesses.

1.3 DEFINITION OF SUPPLY CHAIN

A supply chain is the group of people, companies, resources, and technology involved in the production, transportation, sale, or delivery of a goods or services. In other words, a supply chain is a group of organizations that is linked by a common goal. Their individual roles can be to bring a product or service downstream to the retailer/end customer or upstream to the original supplier.

Every business, including those that provide services, has a supply chain. Before acquiring open source to begin building that application, a software provider, for example, must first decide which product to develop next based on user demand. After that, it must evaluate the application's quality, address any bugs or security flaws, create a marketing strategy, expand an advertising plan, and establish sales and distribution channels. The seller can also think about which other systems application needs to integrate with and work and collaborate with that software provider who builds those connectors.

A supply chain network, in essence, includes all of the procedures involved in transforming subcomponents and materials into the desired finished product. Supply chain management (SCM) is all about managing that network and figuring out how to make it better.

1.4 DEFINITION OF SUPPLY CHAIN MANAGEMENT

SCM refers to the entire process of converting a raw material into finished goods or services, as well as the labor required to transport and deliver those goods or services to associates and, eventually, customers. Supply chain management is the activity of coordinating sourcing, production, inventory management, and transportation among all participants in a supply chain to improve efficiency and customer satisfaction. Many businesses save time and money by assessing and optimizing supply chain management.

SCM refers back to the making plans phase, in addition to the transmission and control of data and capital that happens at some point of the supply chain. Planning, Procurement, Manufacturing, Delivery, and Return are the 5 main components of this discipline.

1.5 SUSTAINABLE SUPPLY CHAIN MANAGEMENT?

Sustainable supply chain management adds the aims of respecting environmental and societal values to traditional supply chain management, which focuses on speed, cost, and dependability of operations. This includes dealing with global issues like weather change, water safety, deforestation, human rights, fair work practices, and sleaze.

1.6 WHAT CONSTITUTES A SUSTAINABLE SUPPLY CHAIN?

Globally, businesses have improved working conditions, cut waste, and reduced carbon emissions. By examining sustainability indicators in supply chain management (SCM) systems, they keep an eye on multi-faceted operations including emphasizing renewable energy, recycling goods and resources, and encouraging more social responsibility among suppliers. The first zero-waste factory in the United States, the Subaru vehicle plant in Indiana, is cited by Scientific American as a persuasive example of a sustainable supply chain.

Companies can also utilize intelligence and predefined procedures to guarantee that things are not transported unnecessarily, such as ensuring that products are shipped from the closest distribution center rather than one on the opposite side of the country.

1.7 IMPORTANCE OF SUPPLY CHAIN MANAGEMENT

Supply chain management is high on their priority list as business leaders recognize the importance of an effective and resilient supply chain. The supply chain is one of your organization's largest expenditures, so it is a good idea to evaluate and improve each cycle of getting an item or service to an end user. This creates loyalty and differentiates the company from its competitor's overtime.

A well-executed supply chain, advanced businesses recognize, reduces waste by keeping supply in sync with demand. Overstock or out-of-stock situations can be avoided when all stakeholders, from supplier to manufacturer to retailer to planner, are on the same page, resulting in wasteful costs, consumer frustration, and lost revenues.

1.8 ROLE OF SUSTAINABILITY IN SUPPLY CHAIN

Mainly in organizations, the supply chain is answerable for the majority of their natural effect. By their actual nature, supply chains often include energy-escalated creation and transportation as merchandise are made and moved all over the planet.

Therefore, associations can frequently have the greatest effect by making changes to their production network instead of other business activities. Supply chain sustainability is also complex due to the numerous supplier relationships and border crossings. This complication can make it difficult to see important operational details like work condition at supplier factory thousands of miles away.

1.9 HOW SUPPLY CHAIN MANAGEMENT WORKS

Supply chain management (SCM) entails connecting all of the organizations that comprise a supply chain for a specific product or service. The majority of items are not purchased, designed, and manufactured by a single company; rather, multiple companies collaborate to create and sell a finished product or service.

1.10 MATERIAL FLOW

Organizations that sell items should figure out the most efficient way to obtain resources, part, or complete goods and then shift them onto the next phase in the chain.

Commodity material flow begins with a supplier procuring raw materials, which then pass through producers, distributors, seller, and, lastly, the end customer; however, certain supply chains may bypass or condense some of these processes. Physical items should be transported to the next location and are frequently stored for a period of time at each step, require planning and skills.

1.11 INFORMATION FLOW

All stakeholders require visibility into detailed status information as commodities or services move through the supply chain. The supplier uses purchase orders from manufacturers to plan assembly runs and assign inventory. A distributor must be aware of the product types and quantities that will be included in upcoming shipments, as well as the estimated arrival dates. A smooth supply chain necessitates a continuous flow of data between all parties involved, including updates on delays, shortages, and other changes. All of this data must be tracked and shared using supply chain management software.

1.12 ADVANTAGES OF SUPPLY CHAIN MANAGEMENT

An effective supply chain management (SCM) strategy has a slew of advantages that make it well worth the investment in people, training, and technology. The subsequent are the main advantages of supply chain managements (SCM):

- **Saving money:** The overriding goal of supply chain management (SCM) is to increase efficiency, which is reflected in cost savings. Companies that can accurately predict demand do not overspend on inventory. Cash flow improves as less money is spent on warehoused products, and businesses benefit from lower production, delivery, and inventory costs. By implementing more efficient processes, businesses can increase productivity while decreasing worker costs. Improved consumer knowledge: Supply chain management (SCM) helps a company avoid running out of a popular product. This keeps customers satisfied and may persuade them to purchase from you rather than a competitor in the future. As companies streamline their supply chains, they may find ways to ship products faster and cheaper, potentially saving money. Finally, increased transparency allows customers to track the status of their orders at any time.
- **Quality issue:** The main goal of SCM is to improve the quality of goods and services. Businesses can develop better quality assurance (QA) processes or discover that a particular supplier or carrier has a high percentage of damaged shipments and work with partners to resolve the issue. Customers will notice tangible quality improvements. Service teams will also appreciate the opportunity to recoup the time and money previously spent on resolving quality-related issues. Effective SCM reduces the number of product recalls and lawsuits that ultimately damage brand reputation.
- **Robust and resilient supply chains:** companies that understand SCM have full visibility into their value chains and can share data with all stakeholders. CFOs can estimate the financial impact if a supplier runs out of a critical component, or if a retailer experiences a decline in sales of a particular product. It is important for the company to be aware of the situation in advance. This allows you to keep your partners informed, reacts quickly, and prevents lost sales or failed customer orders. Companies that have a clear SCM strategy in place can switch quickly in the event of a natural

disaster, disease outbreak, economic uncertainty, or other similar event impacting their supply chain.

- **Better sustainability:** A supply chain that optimizes purchasing and manufacturing reduces waste. Accurate demand planning allows buyers to procure only what they need, reducing inventory wastage. Mastering supply chain sustainability will result in less unnecessary items being manufactured and shipped. This reduces the company's environmental footprint, which is becoming increasingly important to customers. According to Nielsen's Sustainability Survey, 81% of consumers worldwide believe that it is very or very important for companies to limit their environmental impact.
- **Creating a competitive advantage:** The ultimate goal of the SCM initiative is to provide the organization with a significant competitive advantage. Reducing costs, providing a superior customer experience, building a more resilient supply chain, or all of these combined are just some of the benefits that can differentiate a company from its competitors. And when a company reaches or approaches the top, it gains market share and profits rise.

1.13 BENEFITS OF SUPPLY CHAIN SUSTAINABILITY

Supply chain sustainability benefits society and the environment as a whole, not just businesses and their stakeholders. Companies have discovered that, as a result of climate change, catastrophic weather disruptions and increased resource scarcity, for example, can jeopardize their business continuity. Here are five frequently mentioned business activities that benefit from sustainability: Energy costs are decreasing in supply chain operations, for example, if corporations set emission targets with suppliers and assist them in identifying prospective areas for development.

Consumers are more concerned than ever before about where and how products are manufactured. According to researchers at MIT's Sloan School of Management, consumers may spend 2–10% more for items that provide supply chain transparency.

Unsustainable supply chain operation poses an important reputational risk to institutional investors. In recent years, many irresponsible supply chain practices have been reported in the media, and in some cases, this has harmed a company's stock price. Businesses that imported electronics from overseas, maintained unsafe working conditions, used suppliers who routinely polluted local rivers, and purchased defective components or toxic chemicals were all exposed in these accounts. It is also worth noting that nearly half of investors expressed interest in sustainable investing funds in a recent Gallup poll.

Millennials, in particular, demand a stronger sense of purpose in their jobs, according to demographers. Employee hiring and retention are typically influenced by a company's corporate culture and principles, with sustainability playing a significant role.

Governments all over the world are requiring more supply chain sustainability, in part to fulfil the United Nations' 2030 deadline for meeting Sustainable Development Goals such as clean water for everyone. Government laws encompass a wide range of topics, including pharmaceutical traceability, electronic disposal, and conflict mineral avoidance.

1.14 FIVE COMPONENTS OF SUPPLY CHAIN MANAGEMENT (SCM)

SCM is a wide term that refers to an assortment of activities classified into five categories: planning, sourcing, production, delivery, and returns.

- **Planning:** The first step in supply chain management is planning (SCM). A company must first calculate the number of supplies or items required, which is typically done with supply chain and inventory management software, which aids in forecasting accuracy and provides extensive rational. As part of this method, SCM should determine how much labor, money, and partners will be required to meet predicted demand.
- **Sourcing:** Sourcing is the process by which a company determines which suppliers, manufacturers, and distributors are capable of supplying the goods or services specified in its strategy. A manufacturer may have multiple suppliers for a key component to improve supply chain resilience if one source goes out of business or has limited capacity. This adds redundancy, but it may also add complexity. This section of SCM also covers the management of redundant relationships. Manufacturing is an important stage for organizations that outsource manufacturing or purchase fully or partially manufactured items. The company must buy finished goods or all of the parts and materials needed to manufacture items in accordance with its demand

plan, inspect them for quality, and then package the item for direct delivery or allotment.

- **Dispatch:** Dispatch is the last stage of the forward supply chain, and it entails delivering goods or services to customers, who can be other businesses or consumers. To meet promised delivery dates and avoid downstream issues, such as out-of-stock high-demand SKUs, the company must organize and prioritize orders. This component also includes invoice and collects expenditure from consumers.
- **Returns:** The life of a product does not always end when it is delivered to the final user. Customers can return items for a variety of reasons, including dissatisfaction, flaws, surplus of stock, or a warranty claim. The item is then routed back through the supply chain until it reaches the corporation that will issue a refund or replacement. The corporation then scraps, repairs, or returns the object to available inventory.

1.15 SUPPLY CHAIN MANAGEMENT (SCM) PROCESSES

Supply chain management (SCM) encompasses a variety of processes because it covers everything from raw material procurement to final delivery and returns.

- **Purchase:** A company must buy materials or goods before it can manufacture or sell anything. This is where purchasing, also known as procurement, comes in based on the demand plan, companies send purchase orders to their suppliers so that they can meet their quantity and delivery time requirements.
- **Manufacture:** If your company produces goods, this process is an important part of SCM. Manufacturers must plan their production and ensure that they have sufficient capacity to meet demand. Also, you need to keep track of both in-progress and completed tasks.
- **Inventories organization:** A company must monitor the status of all components and items as they are received, used, and shipped out. It must be aware of all inventories entering and exiting its facility, as well as their specifics and location. Inventory management is concerned with increasing inventory turns and lowering holding costs by optimizing stock levels.

- **Order managements:** Companies, particularly merchants, may receive orders from a variety of sources, necessitating the use of a system to organize them. Order management includes sorting and ranking orders as well as directing them to the appropriate fulfillment destinations, such as a provincial storehouse or retail outlet. Order management software also monitors orders as they are packed and deliver to customers.
- **Storehouse managing:** Warehouse management, like inventory management, is concerned with the movement of items within the warehouse as well as the associated workflows. One example is increasing efficiency by direct employees as they put away received product and fulfil client orders by picking, packing, and shipping.
- **Consumer services:** Consumers need to be informed about the status of their custom orders in the supply chain. Giving customers a way to track orders in real time begins with informing them whether you can meet their needs. A high-performing supply chain requires constant communication with partners and customers.
- **Reverse logistic:** According to experts, customers return 30% of their online purchases on average, with some clothing brands having a 50% return rate. Companies must develop efficient methods for dealing with returned goods. This process could be initiated by a supply chain partner, such as a retailer or distributor, or by the end customer. SCM teams should have a clear policy in place for returns, as well as rules outlining when goods should be returned to inventory, repaired, or destroyed.
- **Supply chain performance track:** Leaders need to create supply chain KPIs and other metrics to determine whether their organizations are meeting operational performance standards. To ensure that no issues slip through the cracks, these metrics ought to be monitored in real time and reviewed on a regular basis. Businesses can track output, expenses, fill rate, on-time release rate, and consumer fulfillment.

1.16 ABOUT SCM

In 1982, Keith Oliver, an expert at Booz Allen Hamilton, coined the phrase "supply chain management." The following year, a German corporation launches one of the primary SCM projects, but it would be at least another decade before this method captured the awareness of most business leaders.

Many businesses began examining their supply chains in the mid-to-late 1990s in search of ways to cut costs and improve operations. They started by integrating previously disparate supply chain processes such as planning, manufacturing, and delivery, as well as looking into ways to share data with other businesses in their networks.

Supply chain management (SCM) became more important and widely used as long-distance, global supply chains became the norm over the next few decades. These value chains introduced new risks and challenges, and only careful planning and coordination among an increasing number of partners could keep them running smoothly.

Now, as they explore for new methods to improve their supply chains, major companies are looking into emerging technology.

1.17 SUPPLY CHAIN SUSTAINABILITY BEST PRACTICES

For many businesses, sustainability is no longer an afterthought; it is now an essential component of their supply chain. The Carbon Disclosure Project (CDP) states that its members incorporate environmental data into procurement tools and processes, evaluating suppliers based on environmental criteria in addition to cost and quality metrics.

In contracts and tender agreements, other CDP members incorporate particular environmental performance wording. To aid the cause, some companies provide sustainability training to their procurement managers and suppliers.

1.18 STAGES FOR SUSTAINABLE SUPPLY CHAIN

Forrester Research has developed a model for integrating sustainability into your supply chain that focuses on four key areas:

- **Procuring:** Most businesses begin their sustainability initiatives by focusing on energy and water procurement, as well as sourcing sustainable materials for their goods.
- **Operation:** Look for operational processes or supply chain steps that can be enhanced to save time and capital.

- **Obscurity:** Design things that can be up cycled and reused to reduce waste and obsolete items.
- **Information and communications:** can assess the efficacy of the first three activities and reinforce any relevant actions by engaging with customers, supply chain partners, and other stakeholders.

1.19 SUPPLY CHAIN SUSTAINABILITY TREND

Supply chain sustainability will be driven by innovation in the coming years. Two trends that will keep the momentum going in this area are the circular economy and data-driven supply chains.

A circular economy's purpose is to eliminate pollution and waste from products and systems such as supply chains. Products are designed under this system to have a low environmental impact, survive longer, and be easily disassembled, recreated, or modified for reuse.

In data-driven supply chains, companies use cloud-based supply chain execution systems that provide precise analytics and may even collect data from equipment sensors and other Internet of Things (IoT) devices. As a result of having a better understanding of their operations, businesses can then find, implement, and evaluate methods to become more sustainable.

1.20 WHAT IS SUPPLY CHAIN PLANNING?

Supply chain planning refers to the proactive process of coordinating assets in order to optimize the delivery of goods, services, and information from dealer to consumer while balancing supply and demand (SCP). By utilizing forecasting, pricing strategies, and inventory management strategies, supply chain planning assists you in aligning product supply with consumer demand.

1.21 SUPPLY CHAIN PLANNING VS. SUPPLY CHAIN EXECUTION

It is expected of long-term forecasters and supply chain planners to project outcomes months or even years in advance. However, they do not forecast events based on what they observe in crystal balls. Instead, they base their choices on data from inventory, logistics, and manufacturing.

Supply chain implementation, on the other hand, is the commonplace execution of that plan, including order fulfillment, product transportation, and warehousing. Consider them to be opposite sides of the same coin.

1.22 ABOUT SOFTWARE OF SUPPLY CHAIN MANAGEMENT

Planning and execution are aided by supply chain management software. It estimates demand and controls inventory, allowing you to cut expenses and deliver products more quickly. It gives you real-time data on your supply chain's status, allows you to inspect your operations, and helps you balance supply and insist.

1.23 SIGNIFICANCE OF SUPPLY CHAIN PLAN

Supply chain strategy affects both your top and bottom appearance. Stock shortage can lead to lost sales or increased shipping costs. Too much inventory wastes money. Overestimation and order cancellation can also harm important supplier relationships and disrupt your own business.

1.24 SUPPLY CHAIN PLANNING PROCESS

Strategic, long-term supply chain planning is required to achieve desired company benefits. Managers in a flexible supply chain, on the other hand, have the ability to respond strategically to changing conditions and change tactics on the wing.

1.25 TACTICAL PLAN

This phase establishes a structure for your supply chain and is long-term in nature. This step involves thinking about things like your company's location and suppliers. You will plan transportation routes and sizes, as well as find warehouse space for raw materials and finished goods. If you already have a venue, now is the time to figure out how to maximize its potential.

The following phase looks more deeply into certain components of the framework. You could, for example, set up standing purchase orders with specific suppliers for just-in-time delivery. Even the best-laid plans may necessitate tactical adjustments, such as adjusting sourcing strategies or rebalancing stock among abroad warehouse if, for example, new tariffs increase your expenses.

1.26 SUPPLY CHAIN STRATEGY AND METHOD

Just-in-time delivery and safety stock are frequently combined by planners to lower costs and prepare for shortages. Just before they are needed for manufacturing or other activities, materials are delivered. Less inventories means less money and room.

If there are delays in shipments or material problems, businesses might experience production halts. On the other hand, businesses that employ the safety stock method maintain more inventory as an additional safeguard against supply chain disruptions; however, inventory costs increase as a result.

1.27 SUPPLY CHAIN PLANNING ELEMENTS

Every stage of a product's life cycle necessitates supply chain planning. The following are the steps to take:

Demand forecasting and management: Accurately anticipating demand for a product increases the likelihood of manufacturing and stocking enough inventories to meet customer needs on time, without the need to keep surpluses. Supply chain planners take into account past customer behavior, expected vs. actual sales; promote environment, and other factors. Predictive analytics is increasingly being used to determine how much, where, and when a specific product will be in demand. Supply chain planners then estimate the internal needs of each link in the supply chain.

- **Supply management:** entails locating and acquiring reliable source of raw material, machinery, software, and other commodities used in the production of your product or service.
- **Production managing:** Capacity planning, on the other hand, focuses on your company's assembly line in terms of equipment, personnel, and efficiency.

- **Inventories managing:** Costs, operational efficiency, and customer happiness can all be affected by how a company orders, manages, and stores goods. In order to maintain an acceptable quantity of inventory on hand and maintain a constant level of customer service, businesses work to monitor inventory levels with their supply chain partners.

  Two more financial benefits of inventory management are the ability to benefit from volume discounts from suppliers and correctly account for the asset value of items on corporation balance sheets and tax filings. More advanced inventory management systems can start reorders when stock levels are low in addition to identifying high- and low-performing products.
- **Pricing policy:** Setting the best pricing is a combination of art and science. To manage the supply and demand balance, however, it is necessary to develop an effective strategy. Price reductions, for example, can help you pay the set expenses of assembly lines and your personnel during periods of low demand.
- **Event managing:** Identifying all potential supply chain bottle-necks, failures, and delays at each link in the chain, as well as devising contingency measures such as workarounds, buffers, and backups, is part of this process.
- **Incorporated business planning (IBP):** This is a method of connecting supply chain planning to the rest of your company, from sales to operations to finance. IBP can help firms undertake better predictive analysis to grow their business and manage risk because it collects data from across the enterprise.

1.28 POINT LEADS TO SUCCESS SUPPLY CHAIN

Leading supply chain planners use the following seven essential principles:

Plan and execute the supply chain in concert with the frequency of data updates, planning horizons, and the amount of time your company will project into the future.

Manage data in a systematic way so that you can utilize the same data throughout your supply chain plan and managing procedures. This comprises information about products, customers, manufacturing resources, and suppliers.

Develop cooperative relationships with your major suppliers and customers so you may more readily change predictions, orders, and sales. Look for faults in the data you utilized in your demand planning, such as forecast inaccuracies or bias. Focus on actual point-of-sale data rather than sales orders from your company. Analyze the product lifecycle on a regular basis, searching for ways to improve the relationships between product development and supply chain management. To better understand consumer behavior, cultivate a culture of continuous improvement.

1.29 SUPPLY CHAIN PLANNING TENDENCY

Despite the information that businesses are increasingly relying on technology to better supply chain planning and management, demand forecasting remains a significant barrier for many businesses.

Businesses use predictive analytics to better understand consumer behavior and other demand-creating factors. Planning the production, distribution, and other supply chain services necessary to meet that demand is also becoming increasingly aided by predictive analytics.

KEYWORDS

- **sustainable**
- **environmental**
- **mining**
- **industries**
- **economic**
- **organizations**

CHAPTER 2

Literature Review

ABSTRACT

Sustainability supply chain management in the mining sector involves promoting the health and safety of workers. This includes providing adequate training, protective equipment, and ensuring fair labor practices. By prioritizing the well-being of workers, mining companies can create a more sustainable and ethical supply chain. Overall, sustainability supply chain management in the mining sector is essential for ensuring responsible sourcing, reducing environmental impact, and promoting the well-being of workers. By implementing sustainable practices throughout the supply chain, mining companies can contribute to a more sustainable and ethical industry.

2.1 LITERATURE

As the problem of environmental sustainability becomes an important occupation factor, companies are now looking for decision-making tools to evaluate the most entire image of ecological impact associated with their developed and operation activities. Adhitya et al. (2011) have proposed a frame for green supply chain management by integrating a dynamic SC simulation and LCA indicators to assess the financial and ecological impacts of various SC decisions such as inventory, distribution network configuration, and policy control. Agi and Nishant (2017) have studied a set of 19 factors of influence on the implementation of Green Purchasing Chain Practice Management (GSCM) and analyzed the

Sustainable Supply Chain Management in the Mining Industry.
Suchismita Satapathy and Hullash Chauhan (Authors)

relations between these factors and their effect on the implementation of GSCM practices by the modeling technique. Ahmad et al. (2017) have identified the gap by quantitatively assessing the importance of these forces to SSCM practices in the industry. And they focus on six external forces, that is, financial strength, political strength, stakeholder pressure, competition, energy transition, and regulation. Harja and Helo (2015) have analyzed the environmental impacts such as GHC emissions have been included in supply chain management as additional parameters to traditional costs, lead times, and on-time deliveries. Mirzapour et al. (2013) have developed a stochastic planning approach to solve a multi-site, multi-product, multi-period global productions planning problem in a green supply chain for a medium-term planning horizon under the assumption of production uncertainty demand. Ameknassi et al. (2016) have proposed a programming model, which combines logistics outsourcing decisions with some levels of strategic food chain planning, such as supplies security, customer segmentation, and the liability of the extended manufacturer. The objective is to decrease both the estimated logistics cost and the greenhouse gas (GHG) emission of the supply chain (SC) group, in the context of the uncertainty of the commercial context. Arnette et al. (2014) have introduced a design for sustainability (DFS) taxonomy that is presented to organize and consolidate current technique within these categories. A new DFX concept is being developed that incorporates remanufacture, reuse, and recycling as an environmentally friendly approach to end-of-life. A strategy and life cycle phase framework is being developed to enhance the function of DFX technique by practitioners and to allow for the study of DFX strategies. Babazadeh et al. (2017) have presented the document that shows a multi-objective programming model with multiple governments to design a network of the on-board chain of the second generation's biodiesel supply chain. Future model minimizes the total cost of the biodiesel supply chain from raw material supply centers to customer centers and additionally minimizes the environmental impact (EI) of all processes involved from a well-to-wheel perspective. The mineral industry stands in every industrial sector some of the most difficult sustainability challenges. To ensure its continued "social license" to operate, the industry must respond to these challenges by engaging with its various stakeholders and addressing their sustainability concerns. Industry also needs to be able to measure and benchmark its sustainability performance and demonstrate long-term continuous improvement completed. The mineral sector has already begun to respond to some of the challenges of sustainability occupied by the minerals, minerals and the Sustainable Development Project (MMSD). Azapagic (2004) have suggested contributing to the sectoral activities through the development of a framework for sustainability indicator as an instrument for performance evaluation and improvements. Muduli et al. (2017) have introduced the administration of the Green Supply Chain (GSCM) that focuses on reducing the negative effects of supply chain activities as well as minimizing energy and use of material. Successful implementation of GSCM is important for industries to increase their economic and environmental performance. However, the adoption rate of GSCM in Indian companies, especially in the mining sector, is not encouraging. Barve and Muduli (2013) have introduced the administration of the Green Supply Chain (GSCM) that has received the growing attention in recent years. Van Berkel (2007) has focused on the resource sector; it is assumed that the global search for a sustainable development is one of the most important coaches of future development and business success. Although many uncertainties remain about the full ramifications of sustainable development, it is undisputed that preventative ecological and resource productivity, or eco-efficiency (EE) strategies, depend on the ability of the mining industry to progress in sustainable development. Korol (2013) has studied the assessment of the life cycle (LCA) of steel production through the production of integrated steel and herds for the electric arc furnace in Poland. In addition, it defines the main causes of environmental effect and suggests a strategy to prevent impurities on the majority of the most ecologically friendly steel manufacturing. Biswal et al. (2018) have found the significance of sustainable practices in traditional supply chain management; an effort has been made to build a framework for the study of SSCM strategies using AHP TOPSIS. Bhool and Narwal (2013) have identified critical and important drivers for the implementation of green supply chain management in Indian manufacturing industry in different sectors as well as in two wheeler, four wheeler, and general manufacturing sector companies. Diabat and Govindan (2011) have developed an interpretive structural modeling (ISM) framework; we have developed a model of the drivers impacting the implementation of green supply chain management. The various wheels of administration of the green supply chain (GSCM) are identified according to the GSM literature and consultation with industry experts. Dobele et al. (2014) have examined managing stakeholder relations and communication, and identifying the challenges related to the implementation of corporate social responsibility (CSR) initiatives. Eltayeb et al. (2011) have studied to evaluate the actual environmental, economic, and intangible results resulting from the assumption of the initiatives of the green supply chain. Franks et al. (2011) have examined the minerals industries' responses to sustainable development in the area of waste management, arguing that leadership and management are still needed to forge collective agreements on norms and standards of practice. And they develop set of principles of sustainable development to eliminate mineral and mineral processing waste and analyze the impact on current and future practices. Ghose (2003) has identified that India is one of the first nations of mineral production in the world and its economy depends on the value of the minerals produced. Although mineral productions have enlarged significantly since the country's independence, which remains to be overlooked; it is the contribution of small mines residing at the exit of national minerals. Ghose (2009) has focused on the consumptions of different sources of primary energy and identified that coal continues to be the source of the most important energy in the foreseeable future. Ghose and Roy (2007) have examined the involvement of small-scale mines in national mining production and employment, practices, and Indian policy on small scale. It identifies the disadvantages of the existing government policies and discusses government's role in improving this sector. Manufacturing industries have recently started to adopt the green concept in supply chain management to focus on ecological issues. However, the industry still struggles to identify the obstacles standing in the way of implementing green supply chain management. Govindan et al. (2014) have focused on identifying barriers to implementing green supply chain management (Green SCM) based on sourcing efficiency. Govindarajulu and Daily (2004) have presented a theoretical frame for ecological performance by examining the critical employer and employee factors that influence environmental performance. The model focuses on integrating senior management engagement, worker empowerment, rewards, feedback and ratings, and ecological performance. Hilson and Murck (2000) have examined the sustainability in the context of corporate mining and provided some strategy for mining industries looking to work more sustainably. Hsu et al. (2013) have studied for modern mankind, governments, and environmentally concerned commercial entities; sustainability and environmental issues are among the most important concerns. Green supply chain management has been recognized as a significant factor in promoting organizational sustainability. Green supply chain management is becoming a main strategy for emerging-market organizations to control their environmental responsibilities. Jabbour and Santos (2008) have analyzed the most important contributions of human resource elements to a company's environmental management. Jenkins and Yakovleva (2006) have studied concerns about business sustainability and social responsibility (CSR) have grown in prominence in many countries and industries, none more so than the mining industry. One result of the CSR agenda for mining is that individual enterprises are increasingly required to justify their existence and prove their performance through the revelation of social and environmental data. India's mining and mineral industries, especially small and medium enterprises (SMEs), are focusing on growing their environmental performance by integrating sustainable practices into their management traditional supply chains. Jia et al. (2015) have analyzed to identify common SSCM practices in the mining and mineral industry using interpretive structural modeling (ISM). The mineral industry can interact with water in a variety of ways, affecting the environment and communities in the circumstance of a new debate about the status of access to water as a human right in itself and the involvement of the mining industry in the human rights debate. Kemp et al. (2010) have focused on the relevance of technical, scientific, and engineering approaches to water management on the one hand and human rights perspectives on the other. Lee (2008) has evaluated to explore green supply chain management practice, government participation, and the internal will of suppliers themselves as possible drivers for buyers. Luthra et al. (2015) have identified and analyzed the key success factors behind achieving environmental sustainability in the Indian automotive industry supply chain. Green Supply Chain Management (GSCM) enables organizations to improve their environmental performance to enhance their social image and improve economic performance through price reduction and price-saving principles. In fact, the success of GSCM can lead to an increase in the organization's contribution to sustainable development (SD) (Muduli and Barve, 2013). Mitra and Datta (2014) have found that the level of adoption of GSCM practices by Indian companies is still in its infancy, consumer awareness of ecological sustainability is very low, and there is a lack of regulatory framework to promote ecological sustainability.

Nikolaou and Evangelinos (2010) have carried out many advantages and challenges for companies adopting environmental management practices, but the Greek mining and minerals industry (GMMI) faces similar challenges. Strengths-weaknesses-opportunities-threats analysis (SWOT) was conducted to analyze the challenges faced by GMMI, investigating the strengths, weaknesses, opportunities, and threats the industry faces when adopting environmental management practices. Norgate and Haque (2010) have found that efforts are made to reduce the increase in greenhouse gas emission from mining and mineral processing, which is anticipated in the future due to lower ore grades and higher ore availability should be grinded. In the mining sectors, the community is a particularly important governance entity. Traditional approaches to mineral development are no longer sufficient for these communities demanding a greater share of profits and greater participation in decision-making (Prno and Slocombe, 2012); these trends have been driven by the growth of the sustainable development paradigm and the changing governments that increasingly delegate governance to non-state actors. Shen et al. (2015) have studied and examined a "reasonable implementation approach" and "continuous improvement" are the weaknesses of GSCM practice in the Indian mining sector. Mining evokes strong emotions, despite the crucial benefits it brings to society. Mining operations, if not carefully planned and controlled, cause unacceptable damage to the environment and ecosystems. Rapid, uncontrolled, and sometimes biased activity leads to air, water and noise pollution, land degradation, health hazards, forest wealth and loss of farmland, well depletion and major environmental degradation causes problems (Singh et al., 2008). Suppen et al. (2006) have presented an overview of the Mexico mining industry from the perspective of technological, historical, and economic development. Mining companies' efforts to address environmental management and sustainability issues are reflected in national and international frameworks, and Mexico's ecological regulatory framework for mining sector is analyzed. Vintró et al. (2014) have identified the application of environmental practice in small and medium enterprises in the surface mining industries in Catalonia (Spain). Walker et al.'s (2008) literature reviews identify key categories of internal and external drivers for green supply chain management practices, such as organizational factors, regulations, customers, competitors, and society, but evidence of suppliers as drivers in green supply chain management is almost nonexistent. Internal barriers include cost and lack of authority, and external barriers include guideline, inadequate supplier involvement, and industry-specific barriers. Wittstruck and Teuteberg (2012) have analyzed the research; organizations are often unable to identify key success factors for sustainable supply chain management (SSCM) and their impact on management operations. Zhang et al. (2009) have investigated environmental management practices of small- and medium-sized enterprises, and identified the drivers and barriers to their involvement in environmental management initiatives. Zhu and Sarkis (2006) have compared green supply chain management (GSCM) drivers and practices in a developing country, China, focusing on three typical sectors, the automotive industry, the thermal power plant, and the electrical/electronics industry. Bhanot et al. (2017) have proposed a complete manufacturing sustainability framework to support enablers and moderate barriers based on response from researchers and industry experts. The decision-making test and assessment laboratory approach was applied to identify the most influential factors among the ten enablers and barriers identified in both groups. Chen et al. (2011) have studied that the turnover intent has been a significant issue for decades, as organization has long documented that low employee turnover intent is beneficial to achieving high managerial performance as a result and avoiding the potential costs of recruit and training new employees. Souza et al. (2011) have analyzed completion barriers to lean thoughts in healthcare, based on the experience of applying lean idea in the UK's National Health Service (NHS). Satapathy (2017) has proposed to develop a new technique in which to determine the interrelationship between the barriers that hamper the implementation of efficient recycling process in the plastics sectors of India. Shibin et al. (2016) have designed a theoretical framework of those enablers and barriers of flexible green supply chain management (FGSCM). This method is unique in that it clearly depicts the enablers and barriers, as well as their complex interrelationships, that influence the design and implementation of flexible and green strategies in a supply chain closed-loop system. Talib et al. (2011) have sought to identify the "driving barriers" (those that drive the other obstacles) and the "dependent barriers" (those that are prejudiced by the other barriers) (i.e., which are subjective by others). Sarkis (2003) has focused on the components and parts of green supply chain management which will be the emphasis of this article, as well as how they serve as a foundation for the decision framework. And it is looked at how a lively nonlinear multi-attribute decision-making model, also recognized as the analytical network process, can be used to make decisions in the green supply chain. Rao (2002) has determined the extent to which this greening of the supply chain is occurring in South East Asia; survey questionnaires were used as the research device in a study that included The Philippines, Indonesia, Malaysia, Thailand, and Singapore (Zhu et al. 2007). Increasing pressures from several angles have prompted Chinese automotive supply chain managers to investigate and implement green supply chain management (GSCM) strategy in order to improve both their economic and ecological performances. Min and Galle (1997) have studied an empirical survey of NAPM member in enterprise with a high degree of awareness and frequent use of "green" purchasing are presented in this article. Environmental elements that may influence supplier's selection decisions are highlighted. Klassen and Laughlin (1993) have proposed the potential link between the idea, approach, methodology, and benefits of total quality management (TQM) and ecological excellence in operation management. It is a hierarchical model for the development of environmental excellence. Ecological management has evolved into a strategic endeavor, necessitating the integration of product design, production, delivery, circulation, use, and disposal in order to attain better ecological performance. Gollop and Roberts (1983) have analyzed the effect of sulfur dioxide emission restrictions on the rate of productivity growth in the electric power industries over the 1973–1979 business cycles. Handfield et al. (2002) have introduced the environmental dimension that is added to purchase decisions; it introduces a new set of trade-offs complex the decision-making process with both qualitative and quantitative aspects. Despite this, few businesses utilize any kind of organized study to evaluate suppliers in terms of environmental factors. The analytical hierarchy process (AHP) is used as a decision support tool in this study to help managers comprehend the trade-offs between environmental dimensions. Keating et al. (2008) have purposed to give industries an academe guidance on how to best address the increasing of building a sustainable supply chain. Wagner et al. (2002) have discussed the relationship between environmental and economic performance of enterprises in the European paper manufacturing industries which is investigated in this article. It starts with a discussion of probable functional linkages between environmental and economic performance based on various theoretical frameworks, and then moves on to current empirical and theoretical evaluations of the Porter hypothesis. Gil et al. (2001) have discussed the factors that verify the implementation of ecological management practices and their effects on the financial performance of companies. The empirical evidence sustaining this analysis comes from the Spanish hotel sector. Szwilski (2000) has studied to improve ecological performance by allowing industries to be creative and apply the most suitable and cost-effective solutions. EMSs are projected to be a catalyst for a paradigm shift in ecological regulatory enforcement. Hanna et al. (2000) have examined the links between process type, operational performance, employee engagement, and ecological performance using data from 349 worker involvement (EI) team projects. Zhu and Sarkis (2007) have investigated the links between green supply chain management (GSCM) practice and ecological and economic performance, while taking into account three moderating factors: sell, regulatory, and competitive institutional forces. Zsidisin and Hendrick (1998) have examined purchasing managers' engagement in environmental concerns in Germany, the United Kingdom, and the United States to the level they believe they should have. Hazardous materials, investment recovery, product design, and supply chain interactions are among the environmental challenges covered in this article. Zsidisin and Siferd (2001) have examined ecological research in the supply chain management literature to provide a framework for present and future growth of environmental purchasing theory. Rao and Holt (2005) have identified the potential links between green supply chain management as an initiative to improve the environment, economic performance, and competitiveness among a sample of Southeast Asian companies. Menon and Menon (1997) have discussed the model of an environmentally friendly corporate marketing strategy that reflects the confluence of social performance goals, corporate directions, and marketing strategy by integrating ecological concerns in the development of marketing policies and practices. Awasthi and Grzybowska (2014) have identified a solution to this problem based on supply chain integration barriers using the DEcision MAking Trial and Evaluation Laboratory (DEMATEL) methodology. Alexander et al. (2014) have studied to help construct the decision theory (DT) concepts in sustainable supply chain management (SSCM) research that is explored. Carter and Easton (2011) have purposed to conduct a systematic review of the sustainable supply chain management (SSCM) literature in leading logistics and supply chain management journals over a 20-year period. Ciliberti et al. (2008) have examined the method used by small- and medium-sized enterprises

(SMEs) to transmit socially responsible behaviors to suppliers in developing nations, as well as the challenges they face. Diabat and Govindan (2011) have developed an interpretive structural modeling (ISM) framework of the drivers impacting the implementation of green supply chain management. The various drivers of green supply chain management (GSCM) are identifying based on the GSM literature and in consultation with industry experts (Cripps et al. 2006). Sustainable supply chain management (SSCM) is a term that describes the trend of using purchasing policies and practices to aid in the development of a tourism destination's long-term sustainability. The majority of study has concentrated on manufacturing's environmental issues, whereas other areas of sustainability or challenges confronting the service sector have been largely overlooked. Gandhi et al. (2015) have studied to assess the significant factors connected with the successful implementation of green supply chain management (GSCM). Industries are being forced to embrace green supply chain management (GSCM) efforts as a result of increasing pressures from stakeholders, government, and non-government entities. For industries to improve economic–environmental performance and maintain corporate sustainability, successful implementation of GSCM is critical. Gardas et al. (2018) have presented the analysis to identify and model the critical challenge to *sustainable development* in the textile and apparel sectors. Govindan et al. (2015) have paper investigated the twelve universal drivers of green manufacturing from the combine support of existing literature, industry managers, and expert opinions in the relevant field. Govindan et al. (2014) have identified procurement-based barriers to the adoption of a green supply chain management (Green SCM). Harland (1996) has discussed the various definitions of supply chain management, summarized the related body of knowledge, and connected them using a system approach. The layers of supply chain management systems are known as the internal supply chain, the external supply chain, and the business-to-business network. Islam et al. (2020) have studied the sustainable supply chain management (SSCM) which has gained a lot of momentum in this competitive trade world for every type of industry. The leather industry (leather, leather products, and footwear) is the second export-oriented sector of Bangladesh that SSCM struggles with in its work cycle. Green supply chain management (GSCM) entails taking into account the environmental effects of all activities in a typical supply chain in order to reduce their negative implications. Kaur et al. (2018) have developed DEMATEL-primarily based totally technique for investigating boundaries in GSCM in Canadian context. Khan et al. (2018) have purposed to identify the dimension of social sustainability drive in healthcare in the United Arab Emirates (UAE). The effects of these characteristics on motivation for social sustainability are also investigated. Koberg and Longoni (2019) have studied to present a synthesis of the fundamental features of global supply chain management that are sustainable. Lintukangas et al. (2016) have investigated the issues primarily concerned with supply risks and green supply management. The study looks into whether the focal firm's purchasing function's ability to minimize various forms of supply risks is linked to the company's adoption of green supply management. Lorek and Spangenberg (2014) have identified the sustainable development remains a compelling notion, avoiding the complexity generated by partisan interests reinterpreting the concept. It can still guide robust sustainable consumption if it focuses on meeting human needs while also respecting environmental boundaries. Luthra et al. (2011) have developed a structural model of the barriers to the implementation of green supply chain management (GSCM) in the Indian automotive industries. Design/Methodology/Approach: We identify several barriers and contextual relationships between the knowing barriers. Sustainable consumption and production (SCP) is becoming a major priority for businesses and organizations seeking to secure long-term viability. However, due to the presence of many barriers, the acceptance and execution of SCP efforts in a supply chain can be a huge difficulty. Luthra et al. (2016) have purposed to identify and assess barriers to the adoption of SCP initiatives in the supply chain. Majumdar and Sinha (2019) have investigated the essential barriers of inexperienced fabric and garb deliver chain control in Southeast Asian countries. Twelve essential barriers were recognized through the literature assessment and questionnaire survey. Mathiyazhagan et al. (2013) have identified the barrier is acting as the most leading one for the adoption of green supply chain management, with the goal of assisting industries in making the adoption of green concepts in their supply chains easier by removing the dominant barrier. Mont et al. (2014) have found the SPREAD Sustainable Lifestyles 2050, a European social platform initiative that brought together academics, European decision-makers, corporations, and civil society to investigate latent sustainable living ideas in 2050. Rajeev et al. (2017) have tried to understand the evolution of sustainability issues by analyzing trends in industries and economies using different methods. Sajjad et al. (2015) have studied to contribute to a better considerate of the motivators for and impediments to SSCM implementation in the New Zealand business context. Sarker et al. (2018) have designed to identify barriers to the implementation of Green Supply Chain Management (GSCM) practices in the footwear industry in Bangladesh. In 2005, the Chinese government launched a broad Energy Saving and Emission Reduction (ESER) programme for sustainable production and consumption in response to rising energy scarcity and environmental concerns, as well as the goal of developing a low-carbon society. Lamjahdi et al. (2020) have proposed a set of key performance indicators for assessing and managing overall performance in the mining industry. Eitiveni et al. (2017) have identified the taxonomy of existing study in the SSCM topic in order to demonstrate the landscape of SSCM literature and suggested taxonomy identifies major characteristics of previous SSCM studies, such as theoretical lenses, Triple Bottom Line dimension, unit of study, context, degree of analysis, research design, outcome, and IS/IT aspects, among others. Consumer pressure to improve the environmental and social impacts of supply chain activities is increasing, forcing organizations to adopt sustainable supply chain practices. The management of a sustainable supply chain, on the other hand, is difficult since it is inter-organizational in nature, comprising multiple stakeholders with distinct and sometimes conflicting objectives and priorities. Based on Affordance Theory, this research seeks to fill this knowledge gap by outlining a set of plausible IT-enabled capabilities (Eitiveni et al., 2018). Fallahpour et al. (2017) have focused on the most significant and applicable criteria, as well as their corresponding subcriteria, for sustainable supplier selection through a questionnaire-based survey. Luciano et al. (2018) have identified through the study lens of globalization contributes to a deeper theoretical and empirical investigation on the sustainable management of the supply chain in developing nations. Gong et al. (2019) have studied to demonstrate that a company's information environment, as measured by customer awareness, has a significant advantageous impact on its sustainability performance and SSCM implementation. Gupta and Palsule (2011) have studied an overview of the current state in the field of sustainable supply chain management. Hou et al. (2019) have recognized the sustainability content of the product and built decision models of the focus and supplier company by using four different strategies in the relation to three pillars of sustainability. Alahyari and Pilevari (2021) have proposed to **evaluate** the sustainability of the supply chain in the automotive sector. Koberg and Longoni (2019) have provided a synthesis of the key components of sustainable supply chain management in global supply chains. And it also contributes to practices by providing global supply chain directions to managers of focal enterprises for enhancing long-term supply chain results. Valinejad and Rahmani (2018) have proposed a complete and trustworthy framework for controlling the supply chain sustainability risks for telecoms firms. Wu et al. (2017) have proposed the key factors or criteria of sustainable supply chain management (SSCM) influencing Indonesian coal companies, using SSCM's "Acceptance, Implementation and Performance" (AIP). Gualandris et al. (2015) have proposed that the sustainable supply chain management has mainly focused on monitoring as a self-governing set of strictly defined assessment activities employed by companies to hold stakeholders accountable. Researchers and supply chain practitioners are becoming increasingly interested in environmental sustainability and green environmental challenges. Luthra et al. (2014) identified and investigated green supply chain management (GSCM) practices in the Indian automotive industry. Six GSCM core practices (with 37 sub-practices) and four expected performance outcomes (with 16 performance) were identified through the implementation of GSCM practices. Zailani et al. (2012) have investigated the extent to which sustainable supply chain management methods are being implemented (environmental purchasing and sustainable packaging). Implementing sustainable supply chain management (SSCM) is an important factor that could push organizations to focus on alleviating environmental issues and providing the economic and social benefits. Roy et al. (2020) have described how various SSCM practices used by apparel manufacturers contribute to improved supply chain environmental and operational performance in modern apparel supply chains. Mani et al. (2016) have studied the economic, environmental, and social aspects of sustainability have been shown to extend beyond organizational boundaries, representing the importance of managing sustainability initiatives throughout the supply chain. Though scholars and practitioners pay close attention to economic and ecological sustainability in supply chains, social dimensions receive less attention. Marshall et al. (2015) have found to demonstrate theoretically sound constructs based on four underlying sustainable supply chain management practices: monitoring, system implementation, new product and process development, and strategy redefinitions. Ali et al. (2020) have examined the ability of different sectors in developing wealth to manage the green supply chain in relation to 4 aspects of environmental practices, namely green procurement, green logistics, green products, and process and framework design. Tonelli et al. (2013) have provided a systemic implication for sustainable manufacturing, supply chain design, sustainability performance capacity and management, and organizational change. Roy et al. (2018) have proposed to conduct a complete analysis of the enormous literature on sustainable supply chain management (SSCM), with the specific goal of conducting a thematic exploration of the literature to explain the major aspects of SSCM development. Póvoa et al. (2018) have examined the trends and directions of applications of OR methods toward achieving a sustainable supply chain. Haghighi et al. (2016) have proposed a novel hybrid BSC-DEA framework for performance assessment in sustainable supply chains. Esfahbodi et al. (2017) have identified this document is in twofold: first, to determine the role of governance in the implementation of SSCM practices; second, to determine whether *supply chain management* (SSCM) practices can be both environmentally helpful and commercially feasible. Carter and Rogers (2008) have developed a framework and proposition in place of a middle theory of sustainable supply chain management (SSCM) and introduced the idea of sustainability in the field of supply chain management to show the relationships between ecological, social, and economic performance. Seuring and Müller (2008) have discussed the specific features of sustainable supply chains as well as limitation of existing research. Ahiand Searcy (2013) have identified and discussed the published definitions of Green Supply Chain Management (GSCM) and Sustainable Supply Chain Management (SSCM). Gopal and Thakkar (2016) have analyzed the sustainable supply chain management (SSC) practices for the Indian automotive industry and classified the critical factors for its successful implementation (Kashmanian 2015). With increasing awareness of the relevance of a company's supply chain to its production, performance, status, and sustainability, a better knowledge of how a company may connect with its suppliers and other important stakeholders to strengthen its sustainability emphasis is required. Tseng et al. (2019) have found that the study on driver or barrier analysis of green supply chain management shows a downward trend, while there is an increasing trend to apply mathematical optimization models to improve supply chain management decision-making. Engert et al. (2016) have purposed to investigate the incorporation of corporate sustainability into strategic management and to provide a framework for related issues based on the existing literature in this research area. Ansari and Kant (2017) have identified that the linear programming problems are the most widely used solution approach to optimize complex problems of sustainable supply chain management (SSCM). Tebaldi et al. (2018) have proposed to look at how innovation and sustainability are integrated into the larger supply chain framework. Jayaram and Avittathur (2015) have focused on India's emerging economy, as well as the unique insights of this qualitative Delphi study based on grounded theory, contribute significantly to theory development and managerial practice in the field of sustainable supply chain management. Geng et al. (2017) have studied to know the relationship between green supply chain management (GSCM) practices and the performance of manufacturing companies in Emerging Asia (ESAs) based on empirical evidence. Campos et al. (2017) have discussed that the framework contributes to understanding the variables affecting supply chain and manufacturing performance, and supports the identification of opportunities to create value through relationship between countries to develop sustainable innovation. Camacho et al. (2019) have presented a systematic literature review of works that address the problem of supply chain network design (SCND). Luthra and Mangla (2018) have identified the key challenges for Industries 4.0 initiatives and analyzed the key challenges in order to prioritize them for effective Industries 4.0 approaches to supply chain sustainability in emerging markets by taking the perspective of the Indian manufacturing industries. Koberg and Longoni (2019) have proposed a synthesis of the key components of sustainable supply chain management in global supply chains. Gimenez and Tachizawa (2012) have studied to provide a thorough review of the governance structures that are utilized to extend sustainability to suppliers. Kazancoglu et al. (2018) have focused on raw milk losses in the supply chain, as the livestock and dairy sectors are one of the most important sectors of the Turkish economy. Luthra and Mangla (2018) have identified and analyzed different strategies for implementing sustainable supply chain management (SSCM) practices in the Indian context. Paulraj (2011) has discussed the impact of firm-specific resources and capabilities on sustainable supply management (SSM) and sustainability performance. Particularly, enviropreneurship and strategic purchasing are acknowledged as firm-specific talents and resources necessary for adopting sustainable supply practices. Roy et al. (2020) have proposed a decision-making approach for evaluating environmental sustainability strategies based on their impact on an organization whole supply chain network by combining a fuzzy cognitive map (FCM) and data envelopment analysis (DEA). Morais and Silvestre (2018) have studied and analyzed why and how focal company executes and manages social sustainability in their supply chains. Sarkis et al. (2011) have found that green supply chain management (GSCM) is gaining popularity in academia and industries. Yu and Tseng (2014) have proposed the social compliance program (SCP) criteria for evaluating and selecting sustainable global suppliers can be implemented in the field of social compliance. Moktadir et al. (2018) have identified such barriers and investigated the causal relationships between them in order to facilitate the effective implementation of sustainable *supply chain management* (SSCM) in the leather processing industries in Bangladesh. Wolf (2011) has provided a consistent and testable model of sustainable supply chain management integration (SSCMI). Taticchi et al. (2013) have studied the knowledge in the field of sustainable supply chains by providing an important literature review on sustainable supply chain performance measurement (SSCPM). Mathivathanan et al. (2018) have discussed the interactions between practices and increased the likelihood of successful implementation of SSCM practices in the automotive industry. Mani et al. (2018) have investigated the social issues that are important to suppliers and to identify measures and dimensions related to social sustainability in emerging economies. Hsu et al. (2017) have discussed that the SMEs need to use their limited resources effectively and prioritize their performance factors in terms of a Balanced Scorecard (BSC) approach when designing their sustainable development. Johnson and Schaltegger (2016) have discussed the importance of sustainability in small- and medium-sized enterprises (SMEs) and should implement sustainability management tools. Pedersen, E. R. (2009) has studied an overview of the anatomy of small- and medium-sized enterprises (SMEs) that attempt to manage corporate social responsibility (CSR) in the supply chain. Burke and Gaughran (2007) have focused on engineering and manufacturing small- and medium-sized enterprises (SMEs), and it describes a regional analysis of a limited number of these businesses who have succeeded in establishing relevant tools and strategies to meet the current ISO 14001 standard. Gold et al. (2010) have studied that the supply management skills are evolving into core competencies of organizations as competition shifts from a business-to-business level to a cross-supply chain level. Dubey et al. (2017) have used the Total Interpretive Structural Modeling (TISM) in sustainable *supply chain management* (SSCM). Cormack et al. (2008) have investigated the relationships between supply chain maturity and performance, with specific references to both the Business Process Orientation Maturity Model and the Reference Model for the operation of the supply chain. Sarkis (2003) has focused on the component and element of green supply chain management and how they serve up the decision-making framework. Walker et al. (2008) have investigated the factors that motivate or discourage organizations from implementing green supply chain management initiatives. Sarkis et al. (2011) have identified that the green supply chain management (GSCM) is gaining popularity in academics and industry. Feng et al. (2017) have developed a systematic study that quantitatively represents the knowledge structure and the intellectual advancements of *corporate social responsibility* (CSR) for *supply chain management* (SCM). Zhu and Sarkis (2006) have evaluated the drivers and practices of green supply chain management (GSCM) in one developing country, China, with a focus on three industries: automotive, thermal power plants, and electronics/electrical industry. Ageron et al. (2012) have studied sustainable business development has traditionally a lot of attention over the past decade due to the significant attention that governments, for-profit and non-profit organizations have given to ecological, social, and corporate responsibility. Wittstruck and Teuteberg, (2012) have identified and analyzed the factors that underlie successful sustainable supply chain management (SSCM). Hong et al. (2018) have investigated the impact of SSCM practices on supply chain (SC) dynamic capability and enterprise performances (including economic, ecological, and social performance). Esfahbodi et al. (2016) have suggested that sustainable supply chain management (SSCM) practices can be both environmentally necessary and profitable. Hutchins and Sutherland (2008) have examined the social impact metrics, indicator, framework, and initiative related to their ability to assess the social sustainability of supply chains.

KEYWORDS

- **sustainable**
- **strategies**
- **implementation**
- **planning**
- **friendly**
- **ecological**

REFERENCES

Adhitya, A.; Halim, I.; Srinivasan, R. Decision Support for Green Supply Chain Operations by Integrating Dynamic Simulation and LCA Indicators: Diaper Case Study. *Environ. Sci. Technol.* **2011,** *45* (23), 10178–10185.

Ageron, B.; Gunasekaran, A.; Spalanzani, A. Sustainable Supply Management: An Empirical Study. *Int. J. Product. Econ.* **2012,** *140* (1), 168–182.

Agi, M. A.; Nishant, R. Understanding Influential Factors on Implementing Green Supply Chain Management Practices: An Interpretive Structural Modelling Analysis. *J. Environ. Manage.* **2017,** *188*, 351–363.

Ahi, P.; Searcy, C. A Comparative Literature Analysis of Definitions for Green and Sustainable Supply Chain Management. *J. Clean. Product.* **2013,** *52*, 329–341.

Ahmad, W. N. K. W.; Rezaei, J.; Sadaghiani, S.; Tavasszy, L. A. Evaluation of the External Forces Affecting the Sustainability of Oil and Gas Supply Chain Using Best Worst Method. *J. Clean. Product.***2017,** *153*, 242–252.

Ala-Harja, H.; Helo, P. Reprint of "Green Supply Chain Decisions–Case-Based Performance Analysis from the Food Industry". *Transp. Res. Part E* **2015,** *74*, 11–21.

Alahyari, M.; Pilevari, N. CO-Active Neuro-Fuzzy Inference System Application in Supply Chain Sustainability Assessment Based on Economic, Social, Environmental, and Governance Pillars. *J. Syst. Manag.* **2021,** *6* (3), 265–287.

Alexander, A.; Walker, H.; Naim, M. Decision Theory in Sustainable Supply Chain Management: A Literature Review. *Supply Chain Manag.: Int. J.* **2014**.

Ali, S. S.; Kaur, R.; Ersöz, F.; Altaf, B.; Basu, A.; Weber, G. W. Measuring Carbon Performance for Sustainable Green Supply Chain Practices: A Developing Country Scenario. *Central Eur. J. Operat. Res.* **2020,** *28* (4), 1389–1416.

Ameknassi, L.; Aït-Kadi, D.; Rezg, N. Integration of Logistics Outsourcing Decisions in a Green Supply Chain Design: A Stochastic Multi-Objective Multi-Period Multi-Product Programming Model. *Int. J. Product. Econ.* **2016,** *182*, 165–184.

Ansari, Z. N.; Kant, R. A State-of-Art Literature Review Reflecting 15 Years of Focus on Sustainable Supply Chain Management. *J. Clean. Product.* **2017,** *142*, 2524–2543.

Arnette, A. N.; Brewer, B. L.; Choal, T. Design for Sustainability (DFS): The Intersection of Supply Chain and Environment. *J. Clean. Product.* **2014,** *83*, 374–390.
Awasthi, A.; Grzybowska, K. Barriers of the Supply Chain Integration Process. In *Logistics Operations, Supply Chain Management and Sustainability*; Springer: Cham, 2014; pp 15–30.
Azapagic, A. Developing a Framework for Sustainable Development Indicators for the Mining and Minerals Industry. *J. Clean. Product.* **2004,** *12* (6), 639–662.
Babazadeh, R.; Razmi, J.; Pishvaee, M. S.; Rabbani, M. A Sustainable Second-Generation Biodiesel Supply Chain Network Design Problem Under Risk. *Omega* **2017,** *66*, 258–277.
Barbosa-Póvoa, A. P.; da Silva, C.; Carvalho, A. Opportunities and Challenges in Sustainable Supply Chain: An Operations Research Perspective. *Eur. J. Operat. Res.* **2018,** *268* (2), 399–431.
Barve, A.; Muduli, K. Modelling the Challenges of Green Supply Chain Management Practices in Indian Mining Industries. *J. Manuf. Technol. Manag.* **2013.**
Bhanot, N.; Rao, P. V.; Deshmukh, S. G. An Integrated Approach for Analysing the Enablers and Barriers of Sustainable Manufacturing. *J. Clean. Product.* **2017,** *142*, 4412–4439.
Bhool, R.; Narwal, M. S. An Analysis of Drivers Affecting the Implementation of Green Supply Chain Management for the Indian Manufacturing Industries. *Int. J. Res. Eng. Technol.* **2013,** *2* (11), 2319–1163.
Biswal, J. N.; Muduli, K.; Satapathy, S.; Tripathy, S. A Framework for Assessment of SSCM Strategies with Respect to Sustainability Performance: An Indian Thermal Power Sector Perspective. *Int. J. Procure. Manag.* **2018,** *11* (4), 455–471.
Burchart-Korol, D. Life Cycle Assessment of Steel Production in Poland: A Case Study. *J. Clean. Product.* **2013,** *54*, 235–243.
Burke, S.; Gaughran, W. F. Developing a Framework for Sustainability Management in Engineering SMEs. *Robot. Comput.-Integr. Manuf.* **2007,** *23* (6), 696–703.
Campos, J. K.; Straube, F.; Wutke, S.; Cardoso, P. A. Creating Value by Sustainable Manufacturing and Supply Chain Management Practices—A Cross-Country Comparison. *Procedia Manuf.* **2017,** *8*, 686–690.
Carter, C. R.; Easton, P. L. Sustainable Supply Chain Management: Evolution and Future Directions. *Int. J. Phys. Distrib. Logist. Manag.* **2011.**
Carter, C. R.; Rogers, D. S. A Framework of Sustainable Supply Chain Management: Moving Toward New Theory. *Int. J. Phys. Distrib. Logist. Manag.* **2008.**
Chen, M. F.; Lin, C. P.; Lien, G. Y. Modelling Job Stress as a Mediating Role in Predicting Turnover Intention. *Serv. Ind. J.* **2011,** *31* (8), 1327–1345.
Ciliberti, F.; Pontrandolfo, P.; Scozzi, B. Investigating Corporate Social Responsibility in Supply Chains: A SME Perspective. *J. Clean. Product.* **2008,** *16* (15), 1579–1588.
Cripps, K.; Font, X.; Tapper, R.; Kornilaki, M. The Challenges of Sustainable Supply Chain Management in a Service Sector: The Tourism Experience. *Bus. Strat. Environ.* **2006,** *17*, 260.
de Souza, L. B.; Pidd, M. Exploring the Barriers to Lean Health Care Implementation. *Public Money Manag.* **2011,** *31* (1), 59–66.
Diabat, A.; Govindan, K. An Analysis of the Drivers Affecting the Implementation of Green Supply Chain Management. *Resour. Conserv. Recycl.* **2011,** *55* (6), 659–667.
Resour. Conserv. Recycl.Dobele, A. R.; Westberg, K.; Steel, M.; Flowers, K. An Examination of Corporate Social Responsibility Implementation and Stakeholder

Engagement: A Case Study in the Australian Mining Industry. *Bus. Strat. Environ.* **2014,** *23* (3), 145–159.
Dubey, R.; Gunasekaran, A.; Papadopoulos, T.; Childe, S. J.; Shibin, K. T.; Wamba, S. F. Sustainable Supply Chain Management: Framework and Further Research Directions. *J. Clean. Product.* **2017,** *142*, 1119–1130.
Eitiveni, I.; Kurnia, S.; Buyya, R. Sustainable Supply Chain Management: Taxonomy, Gaps, and Future Directions, 2017.
Eitiveni, I.; Kurnia, S.; Buyya, R. IT-Enabled Capabilities for Sustainable Supply Chain Management: An Affordance Theory Perspective, 2018.
Eltayeb, T. K.; Zailani, S.; Ramayah, T. Green Supply Chain Initiatives Among Certified Companies in Malaysia and Environmental Sustainability: Investigating the Outcomes. *Resour. Conserv. Recycl.* **2011,** *55* (5), 495–506.
Engert, S.; Rauter, R.; Baumgartner, R. J. Exploring the Integration of Corporate Sustainability into Strategic Management: A Literature Review. *J. Clean. Product.* **2016,** *112*, 2833–2850.
Esfahbodi, A.; Zhang, Y.; Watson, G. Sustainable Supply Chain Management in Emerging Economies: Trade-Offs Between Environmental and Cost Performance. *Int. J. Product. Econ.* **2016,** *181*, 350–366.
Esfahbodi, A.; Zhang, Y.; Watson, G.; Zhang, T. Governance Pressures and Performance Outcomes of Sustainable Supply Chain Management—An Empirical Analysis of UK Manufacturing Industry. *J. Clean. Product.* **2017,** *155*, 66–78.
Fallahpour, A.; Olugu, E. U.; Musa, S. N.; Wong, K. Y.; Noori, S. A Decision Support Model for Sustainable Supplier Selection in Sustainable Supply Chain Management. *Comput. Ind. Eng.* **2017,** *105*, 391–410.
Feng, Y.; Zhu, Q.; Lai, K. H. Corporate Social Responsibility for Supply Chain Management: A Literature Review and Bibliometric Analysis. *J. Clean. Product.* **2017,** *158*, 296–307.
Franks, D. M.; Boger, D. V.; Côte, C. M.; Mulligan, D. R. Sustainable Development Principles for the Disposal of Mining and Mineral Processing Wastes. *Resour. Policy* **2011,** *36* (2), 114–122.
Gandhi, S.; Mangla, S. K.; Kumar, P.; Kumar, D. Evaluating Factors in Implementation of Successful Green Supply Chain Management Using DEMATEL: A Case Study. *Int. Strat. Manag. Rev.* **2015,** 3 (1–2), 96–109.
Gardas, B. B.; Raut, R. D.; Narkhede, B. Modelling the Challenges to Sustainability in the Textile and Apparel (TandA) Sector: A Delphi-DEMATEL Approach. *Sustain. Product. Consumpt.* **2018,** *15*, 96–108.
Geng, R.; Mansouri, S. A.; Aktas, E. The Relationship Between Green Supply Chain Management and Performance: A Meta-Analysis of Empirical Evidences in Asian Emerging Economies. *Int. J. Product. Econ.* **2017,** *183*, 245–258.
Ghose, M. K. Promoting Cleaner Production in the Indian Small-Scale Mining Industry. *J. Clean. Product.* **2003,** *11* (2), 167–174.
Ghose, M. K. Technological Challenges for Boosting Coal Production with Environmental Sustainability. *Environ. Monitor. Assess.* **2009,** *154* (1), 373–381.
Ghose, M. K.; Roy, S. Contribution of Small-Scale Mining to Employment, Development and Sustainability—An Indian Scenario. *Environ. Dev. Sustain.* **2007,** *9* (3), 283–303.
Gil, M. A.; Jiménez, J. B.; Lorente, J. C. An Analysis of Environmental Management, Organizational Context and Performance of Spanish Hotels. *Omega* **2001,** *29* (6), 457–471.

Gimenez, C.; Tachizawa, E. M. Extending Sustainability to Suppliers: A Systematic Literature Review. *Supply Chain Manag.* 2012.

Gold, S.; Seuring, S.; Beske, P. Sustainable Supply Chain Management and Inter-Organizational Resources: A Literature Review. *Corp. Soc. Respons. Environ. Manag.* **2010,** *17* (4), 230–245.

Gollop, F. M.; Roberts, M. J. Environmental Regulations and Productivity Growth: The Case of Fossil-Fueled Electric Power Generation. *J. Polit. Econ.* **1983,** *91* (4), 654–674.

Gómez-Luciano, C. A.; Domínguez, F. R. R.; González-Andrés, F.; De Meneses, B. U. L. Sustainable Supply Chain Management: Contributions of Supplies Markets. *J. Clean. Product.* **2018,** *184*, 311–320.

Gong, M.; Gao, Y.; Koh, L.; Sutcliffe, C.; Cullen, J. The Role of Customer Awareness in Promoting Firm Sustainability and Sustainable Supply Chain Management. *Int. J. Product. Econ.* **2019,** *217*, 88–96.

Gopal, P. R. C.; Thakkar, J. Sustainable Supply Chain Practices: An Empirical Investigation on Indian Automobile Industry. *Product. Plann. Control* **2016,** *27* (1), 49–64.

Govindan, K.; Diabat, A.; Shankar, K. M. Analyzing the Drivers of Green Manufacturing with Fuzzy Approach. *J. Clean. Product.* **2015,** *96*, 182–193.

Govindan, K.; Kaliyan, M.; Kannan, D.; Haq, A. N. Barriers Analysis for Green Supply Chain Management Implementation in Indian Industries Using Analytic Hierarchy Process. *Int. J. Product. Econ.* **2014,** *147*, 555–568.

Govindarajulu, N.; Daily, B. F. Motivating Employees for Environmental Improvement. *Ind. Manag. Data Syst.* **2004**.

Gualandris, J.; Klassen, R. D.; Vachon, S.; Kalchschmidt, M. Sustainable Evaluation and Verification in Supply Chains: Aligning and Leveraging Accountability to Stakeholders. *J. Operat. Manag.* **2015,** *38*, 1–13.

Gupta, S.; Palsule-Desai, O. D. Sustainable Supply Chain Management: Review and Research Opportunities. *IIMB Manag. Rev.* **2011,** *23* (4), 234–245.

Haghighi, S. M.; Torabi, S. A.; Ghasemi, R. An Integrated Approach for Performance Evaluation in Sustainable Supply Chain Networks (with a Case Study). *J. Clean. Product.* **2016,** *137*, 579–597.

Handfield, R.; Walton, S. V.; Sroufe, R.; Melnyk, S. A. Applying Environmental Criteria to Supplier Assessment: A Study in the Application of the Analytical Hierarchy Process. *Eur. J. Operat. Res.* **2002,** *141* (1), 70–87.

Hanna, M. D.; Newman, W. R.; Johnson, P. Linking Operational and Environmental Improvement Through Employee Involvement. *Int. J. Operat. Product. Manag.* **2000**.

Harland, C. M. Supply Chain Management: Relationships, Chains and Networks. *Br. J. Manag.* **1996,** *7*, S63–S80.

Hilson, G.; Murck, B. Sustainable Development in the Mining Industry: Clarifying the Corporate Perspective. *Resour. Policy* **2000,** *26* (4), 227–238.

Hong, J.; Zhang, Y.; Ding, M. Sustainable Supply Chain Management Practices, Supply Chain Dynamic Capabilities, and Enterprise Performance. *J. Clean. Product.* **2018,** *172*, 3508–3519.

Hou, G.; Wang, Y.; Xin, B. A Coordinated Strategy for Sustainable Supply Chain Management with Product Sustainability, Environmental Effect and Social Reputation. *J. Clean. Product.* **2019,** *228*, 1143–1156.

Hsu, C. C.; Tan, K. C.; Zailani, S. H. M.; Jayaraman, V. Supply Chain Drivers That Foster the Development of Green Initiatives in an Emerging Economy. *Int. J. Operat. Product. Manag.* **2013**.

Hsu, C. H.; Chang, A. Y.; Luo, W. Identifying Key Performance Factors for Sustainability Development of SMEs–Integrating QFD and Fuzzy MADM Methods. *J. Clean. Product.* **2017,** *161*, 629–645.

Hutchins, M. J.; Sutherland, J. W. An Exploration of Measures of Social Sustainability and Their Application to Supply Chain Decisions. *J. Clean. Product.* **2008,** *16* (15), 1688–1698.

Islam, M. H.; Sarker, M. R.; Hossain, M. I.; Ali, K.; Noor, K. A. Towards Sustainable Supply Chain Management (SSCM): A Case of Leather Industry. *J. Operat. Strat. Plann.* **2020,** *3* (1), 81–98.

Jabbour, C. J. C.; Santos, F. C. A. Relationships Between Human Resource Dimensions and Environmental Management in Companies: Proposal of a Model. *J. Clean. Product.* **2008,** *16* (1), 51–58.

Jayaram, J.; Avittathur, B. Green Supply Chains: A Perspective from an Emerging Economy. *Int. J. Product. Econ.* **2015,** *164*, 234–244.

Jenkins, H.; Yakovleva, N. Corporate Social Responsibility in the Mining Industry: Exploring Trends in Social and Environmental Disclosure. *J. Clean. Product.* **2006,** *14* (3–4), 271–284.

Jia, P.; Diabat, A.; Mathiyazhagan, K. Analyzing the SSCM Practices in the Mining and Mineral Industry by ISM Approach. *Resour. Policy* **2015,** *46*, 76–85.

Johnson, M. P.; Schaltegger, S. Two Decades of Sustainability Management Tools for SMEs: How Far Have We Come? *J. Small Bus. Manag.* **2016,** *54* (2), 481–505.

Kashmanian, R. M. Building a Sustainable Supply Chain: Key Elements. *Environ. Qual. Manag.* **2015,** *24* (3), 17–41.

Kaur, J.; Sidhu, R.; Awasthi, A.; Chauhan, S.; Goyal, S. A DEMATEL Based Approach for Investigating Barriers in Green Supply Chain Management in Canadian Manufacturing Firms. *Int. J. Product. Res.* **2018,** *56* (1–2), 312–332.

Kazancoglu, Y.; Ozkan-Ozen, Y. D.; Ozbiltekin, M. Minimizing Losses in Milk Supply Chain with Sustainability: An Example from an Emerging Economy. *Resour. Conserv. Recycl.* **2018,** *139*, 270–279.

Keating, B.; Quazi, A.; Kriz, A.; Coltman, T. In Pursuit of a Sustainable Supply Chain: Insights from Westpac Banking Corporation. *Supply Chain Manag.: Int. J.* **2008**.

Kemp, D.; Bond, C. J.; Franks, D. M.; Cote, C. Mining, Water and Human Rights: Making the Connection. *J. Clean. Product.* **2010,** *18* (15), 1553–1562.

Khan, M.; Hussain, M.; Gunasekaran, A.; Ajmal, M. M.; Helo, P. T. Motivators of Social Sustainability in Healthcare Supply Chains in the UAE—Stakeholder Perspective. *Sustain. Product. Consumpt.* **2018,** *14*, 95–104.

Klassen, R. D.; McLaughlin, C. P. TQM and Environmental Excellence in Manufacturing. *Ind. Manag. Data Syst.* **1993**.

Koberg, E.; Longoni, A. A Systematic Review of Sustainable Supply Chain Management in Global Supply Chains. *J. Clean. Product.* **2019,** *207*, 1084–1098.

Lamjahdi, A.; Bouloiz, H.; Gallab, M. Proposed Overall Performance Indicators for Sustainable Manufacturing Management in Mining Industry. In *13ème conference internationale de modelisation, optimisation et simulation* Nov 2020.

Lee, S. Y. Drivers for the Participation of Small and Medium-Sized Suppliers in Green Supply Chain Initiatives. *Supply Chain Manag.: Int. J.* **2008**.
Lintukangas, K.; Kähkönen, A. K.; Ritala, P. Supply Risks as Drivers of Green Supply Management Adoption. *J. Clean. Product.* **2016,** *112*, 1901–1909.
Lorek, S.; Spangenberg, J. H. Sustainable Consumption within a Sustainable Economy–Beyond Green Growth and Green Economies. *J. Clean. Product.* **2014,** *63*, 33–44.
Luthra, S.; Mangla, S. K. Evaluating Challenges to Industry 4.0 Initiatives for Supply Chain Sustainability in Emerging Economies. *Process Saf. Environ. Protect.* **2018,** *117*, 168–179.
Luthra, S.; Mangla, S. K. When Strategies Matter: Adoption of Sustainable Supply Chain Management Practices in an Emerging Economy's Context. *Resour. Conserv. Recycl.* **2018,** *138*, 194–206.
Luthra, S.; Garg, D.; Haleem, A. Empirical Analysis of Green Supply Chain Management Practices in Indian Automobile Industry. *J. Inst. Eng. (India): Ser. C* **2014,** *95* (2), 119–126.
Luthra, S.; Garg, D.; Haleem, A. Critical Success Factors of Green Supply Chain Management for Achieving Sustainability in Indian Automobile Industry. *Product. Plann. Control* **2015,** *26* (5), 339–362.
Luthra, S.; Kumar, V.; Kumar, S.; Haleem, A. Barriers to Implement Green Supply Chain Management in Automobile Industry Using Interpretive Structural Modeling Technique: An Indian Perspective. *J. Ind. Eng. Manag. (JIEM)* **2011,** *4* (2), 231–257.
Luthra, S.; Mangla, S. K.; Xu, L.; Diabat, A. Using AHP to Evaluate Barriers in Adopting Sustainable Consumption and Production Initiatives in a Supply Chain. *Int. J. Product. Econ.* **2016,** *181*, 342–349.
Majumdar, A.; Sinha, S. K. Analyzing the Barriers of Green Textile Supply Chain Management in Southeast Asia Using Interpretive Structural Modeling. *Sustain. Product. Consumpt.* **2019,** *17*, 176–187.
Mani, V.; Gunasekaran, A.; Delgado, C. Enhancing Supply Chain Performance Through Supplier Social Sustainability: An Emerging Economy Perspective. *Int. J. Product. Econ.* **2018,** *195*, 259–272.
Mani, V.; Gunasekaran, A.; Papadopoulos, T.; Hazen, B.; Dubey, R. Supply Chain Social Sustainability for Developing Nations: Evidence from India. *Resour. Conserv. Recycl.* **2016,** *111*, 42–52.
Marshall, D.; McCarthy, L.; Heavey, C.; McGrath, P. Environmental and Social Supply Chain Management Sustainability Practices: Construct Development and Measurement. *Product. Plann. Control* **2015,** *26* (8), 673–690.
Mathivathanan, D.; Kannan, D.; Haq, A. N. Sustainable Supply Chain Management Practices in Indian Automotive Industry: A Multi-Stakeholder View. *Resour. Conserv. Recycl.* **2018,** *128*, 284–305.
Mathiyazhagan, K.; Govindan, K.; NoorulHaq, A.; Geng, Y. An ISM Approach for the Barrier Analysis in Implementing Green Supply Chain Management. *J. Clean. Product.* **2013,** *47*, 283–297.
McCormack, K.; Ladeira, M. B.; de Oliveira, M. P. V. Supply Chain Maturity and Performance in Brazil. *Supply Chain Manag.: Int. J.* **2008**.
Menon, A.; Menon, A. Enviropreneurial Marketing Strategy: The Emergence of Corporate Environmentalism as Market Strategy. *J. Market.* **1997,** *61* (1), 51–67.

Min, H.; Galle, W. P. Green Purchasing Strategies: Trends and Implications. *Int. J. Purchas. Mater. Manag.* **1997,** *33* (2), 10–17.

Mirzapour Al-e-hashem, S. M. J.; Baboli, A.; Sazvar, Z. A Stochastic Aggregate Production Planning Model in a Green Supply Chain: Considering Flexible Lead Times, Nonlinear Purchase and Shortage Cost Functions. *Eur. J. Operat. Res.* **2013,** *230* (1), 26–41.

Mitra, S.; Datta, P. P. Adoption of Green Supply Chain Management Practices and Their Impact on Performance: An Exploratory Study of Indian Manufacturing Firms. *Int. J. Product. Res.* **2014,** *52* (7), 2085–2107.

Moktadir, M. A.; Ali, S. M.; Rajesh, R.; Paul, S. K. Modeling the Interrelationships Among Barriers to Sustainable Supply Chain Management in Leather Industry. *J. Clean. Product.* **2018,** *181*, 631–651.

Mont, O.; Neuvonen, A.; Lähteenoja, S. Sustainable Lifestyles 2050: Stakeholder Visions, Emerging Practices and Future Research. *J. Clean. Product.* **2014,** *63*, 24–32.

Morais, D. O.; Silvestre, B. S. Advancing Social Sustainability in Supply Chain Management: Lessons from Multiple Case Studies in an Emerging Economy. *J. Clean. Product.* **2018,** *199*, 222–235.

Moreno-Camacho, C. A.; Montoya-Torres, J. R.; Jaegler, A.; Gondran, N. Sustainability Metrics for Real Case Applications of the Supply Chain Network Design Problem: A Systematic Literature Review. *J. Clean. Product.* **2019,** *231*, 600–618.

Muduli, K.; Barve, A. Establishment of a Sustainable Development Framework in Small Scale Mining Supply Chains in India. *Int. J. Intell. Enterprise* **2013,** *2* (1), 84–100.

Muduli, K.; Biswal, J. N.; Satapathy, S.; Barve, A.; Tripathy, S. Investigation of Influential Factors of Green Supply Chain Management in Indian Mining Industries: An Empirical Study. *Int. J. Bus. Excellence* **2017,** *12* (3), 351–375.

Nikolaou, I. E.; Evangelinos, K. I. A SWOT Analysis of Environmental Management Practices in Greek Mining and Mineral Industry. *Resour. Policy* **2010,** *35* (3), 226–234.

Norgate, T.; Haque, N. Energy and Greenhouse Gas Impacts of Mining and Mineral Processing Operations. *J. Clean. Product.* **2010,** *18* (3), 266–274.

Paulraj, A. Understanding the Relationships Between Internal Resources and Capabilities, Sustainable Supply Management and Organizational Sustainability. *J. Supply Chain Manag.* **2011,** *47* (1), 19–37.

Pedersen, E. R. The Many and the Few: Rounding up the SMEs That Manage CSR in the Supply Chain. *Supply Chain Manag.: Int. J.* **2009**.

Prno, J.; Slocombe, D. S. Exploring the Origins of 'Social License to Operate' in the Mining Sector: Perspectives from Governance and Sustainability Theories. *Resour. Policy* **2012,** *37* (3), 346–357.

Rajeev, A.; Pati, R. K.; Padhi, S. S.; Govindan, K. Evolution of Sustainability in Supply Chain Management: A Literature Review. *J. Clean. Product.* **2017,** *162*, 299–314.

Rao, P. Greening the Supply Chain: A New Initiative in South East Asia. *Int. J. Operat. Product. Manag.* **2002**.

Rao, P.; Holt, D. Do Green Supply Chains Lead to Competitiveness and Economic Performance. *Int. J. Operat. Product. Manag.* **2005**.

Roy, S.; Das, M.; Ali, S. M.; Raihan, A. S.; Paul, S. K.; Kabir, G. Evaluating Strategies for Environmental Sustainability in a Supply Chain of an Emerging Economy. *J. Clean. Product.* **2020,** *262*, 121389.

Roy, V.; Schoenherr, T.; Charan, P. The Thematic Landscape of Literature in Sustainable Supply Chain Management (SSCM): A Review of the Principal Facets in SSCM Development. *Int. J. Operat. Product. Manag.* **2018**.

Roy, V.; Silvestre, B. S.; Singh, S. Reactive and Proactive Pathways to Sustainable Apparel Supply Chains: Manufacturer's Perspective on Stakeholder Salience and Organizational Learning Toward Responsible Management. *Int. J. Product. Econ.* **2020,** *227*, 107672.

Sajjad, A.; Eweje, G.; Tappin, D. Sustainable Supply Chain Management: Motivators and Barriers. *Bus. Strat. Environ.* **2015,** *24* (7), 643–655.

Sarker, M. R.; Ahmed, F.; Deb, A. K.; Chowdhury, M. Identifying Barriers for Implementing Green Supply Chain Management (GSCM) in Footwear Industry of Bangladesh: A Delphi Study Approach. *Revista de Pielarie Incaltaminte* **2018,** *18* (3), 175.

Sarkis, J. A Strategic Decision Framework for Green Supply Chain Management. *J. Clean. Product.* **2003,** *11* (4), 397–409.

Sarkis, J.; Zhu, Q.; Lai, K. H. An Organizational Theoretic Review of Green Supply Chain Management Literature. *Int. J. Product. Econ.* **2011,** *130* (1), 1–15.

Satapathy, S. An Analysis of Barriers for Plastic Recycling in the Indian Plastic Industry. *Benchmarking: Int. J.* **2017**.

Seuring, S.; Müller, M. From a Literature Review to a Conceptual Framework for Sustainable Supply Chain Management. *J. Clean. Product.* **2008,** *16* (15), 1699–1710.

Shen, L.; Muduli, K.; Barve, A. Developing a Sustainable Development Framework in the Context of Mining Industries: AHP Approach. *Resour. Policy* **2015,** *46*, 15–26.

Shibin, K. T.; Gunasekaran, A.; Papadopoulos, T.; Dubey, R.; Singh, M.; Wamba, S. F. Enablers and Barriers of Flexible Green Supply Chain Management: A Total Interpretive Structural Modeling Approach. *Glob. J. Flexible Syst. Manag.* **2016,** *17* (2), 171–188.

Singh, G.; Shukla, A. K. Environmental Issues with Best Practice Examples of Responsible Mining in India. In *21st World Mining Congress and Expo* **2008, 2008,** Nov; pp 7–12.

Suppen, N.; Carranza, M.; Huerta, M.; Hernández, M. A. Environmental Management and Life Cycle Approaches in the Mexican Mining Industry. *J. Clean. Product.* **2006,** *14* (12–13), 1101–1115.

Szwilski, T. B. Using Environmental Management Systems to Systematically Improve Operational Performance and Environmental Protection. *Int. J. Surf. Mining Reclamation Environ.* **2000,** 14 (3), 183–191.

Talib, F.; Rahman, Z.; Qureshi, M. N. Analysis of Interaction Among the Barriers to Total Quality Management Implementation Using Interpretive Structural Modeling Approach. *Benchmarking: Int. J.* **2011**.

Taticchi, P.; Tonelli, F.; Pasqualino, R. Performance Measurement of Sustainable Supply Chains: A Literature Review and a Research Agenda. *Int. J. Productivity Perform. Manag.* **2013**.

Tebaldi, L.; Bigliardi, B.; Bottani, E. Sustainable Supply Chain and Innovation: A Review of the Recent Literature. *Sustainability* **2018,** *10* (11), 3946.

Tonelli, F.; Evans, S.; Taticchi, P. Industrial Sustainability: Challenges, Perspectives, Actions. *Int. J. Bus. Innov. Res.* **2013,** *7* (2), 143–163.

Tseng, M. L.; Islam, M. S.; Karia, N.; Fauzi, F. A.; Afrin, S. A Literature Review on Green Supply Chain Management: Trends and Future Challenges. *Resour. Conserv. Recycl.* **2019,** *141*, 145–162.

Valinejad, F.; Rahmani, D. Sustainability Risk Management in the Supply Chain of Telecommunication Companies: A Case Study. *J. Clean. Product.* **2018,** *203*, 53–67.

Van Berkel, R. Eco-Efficiency in the Australian Minerals Processing Sector. *J. Clean. Product.* **2007,** *15* (8–9), 772–781.

Vintró, C.; Sanmiquel, L.; Freijo, M. Environmental Sustainability in the Mining Sector: Evidence from Catalan Companies. *J. Clean. Product.* **2014,** *84*, 155–163.

Wagner, M.; Van Phu, N.; Azomahou, T.; Wehrmeyer, W. The Relationship Between the Environmental and Economic Performance of Firms: An Empirical Analysis of the European Paper Industry. *Corp. Soc. Respons. Environ. Manag.* **2002,** *9* (3), 133–146.

Walker, H.; Di Sisto, L.; McBain, D. Drivers and Barriers to Environmental Supply Chain Management Practices: Lessons from the Public and Private Sectors. *J. Purchasing Supply Manag.* **2008,** *14* (1), 69–85.

Wittstruck, D.; Teuteberg, F. Understanding the Success Factors of Sustainable Supply Chain Management: Empirical Evidence from the Electrics and Electronics Industry. *Corp. Soc. Respons. Environ. Manag.* **2012,** *19* (3), 141–158.

Wolf, J. Sustainable Supply Chain Management Integration: A Qualitative Analysis of the German Manufacturing Industry. *J. Bus. Ethics* **2011,** *102* (2), 221–235.

Wu, J. Z.; Santoso, C. H.; Roan, J. Key Factors for Truly Sustainable Supply Chain Management: An Investigation of the Coal Industry in Indonesia. *Int. J. Logist. Manag.* **2017.**

Yu, V. F.; Tseng, L. C. Measuring Social Compliance Performance in the Global Sustainable Supply Chain: An AHP Approach. *J. Inf. Optim. Sci.* **2014,** *35* (1), 47–72.

Zailani, S.; Jeyaraman, K.; Vengadasan, G.; Premkumar, R. Sustainable Supply Chain Management (SSCM) in Malaysia: A Survey. *Int. J. Product. Econ.* **2012,** *140* (1), 330–340.

Zhang, B.; Bi, J.; Liu, B. Drivers and Barriers to Engage Enterprises in Environmental Management Initiatives in Suzhou Industrial Park, China. *Front. Environ. Sci. Eng. China* **2009,** *3* (2), 210–220.

Zhu, Q.; Geng, Y. Drivers and Barriers of Extended Supply Chain Practices for Energy Saving and Emission Reduction Among Chinese Manufacturers. *J. Clean. Product.* **2013,** *40*, 6–12.

Zhu, Q.; Sarkis, J. An Inter-Sectoral Comparison of Green Supply Chain Management in China: Drivers and Practices. *J. Clean. Product.* **2006,** *14* (5), 472–486.

Zhu, Q.; Sarkis, J. The Moderating Effects of Institutional Pressures on Emergent Green Supply Chain Practices and Performance. *Int. J. Product. Res.* **2007,** *45* (18–19), 4333–4355.

Zhu, Q.; Sarkis, J.; Lai, K. H. Green Supply Chain Management: Pressures, Practices and Performance Within the Chinese Automobile Industry. *J. Clean. Product.* **2007,** *15* (11–12), 1041–1052.

Zsidisin, G. A.; Hendrick, T. E. Purchasing's Involvement in Environmental Issues: A Multi-Country Perspective. *Ind. Manag. Data Syst.* **1998.**

Zsidisin, G. A.; Siferd, S. P. Environmental Purchasing: A Framework for Theory Development. *Eur. J. Purchas. Supply Manag.* **2001,** *7* (1), 61–73.

CHAPTER 3

A Framework for Sustainable Challenges of the Indian Mining Sector

ABSTRACT

The significance of sustainable supply chain management (SSCM) in present years has been known by various groups of both educators and researchers. Further, more consumer awareness in developing countries of how important it is to pay attention to ecological and social influences also require companies to be aware of their supply chain. Moreover, several studies were carried out on the introduction of SSCM in the form of simulation and its functioning in various industrial sectors. Although a study of the overall SSCM implementation still focused on developing countries, there have been no focus on the circumstances in each country. To overcome this gap, this research has been conducted to categorize the criteria for the selection of sustainable suppliers in the SSCM of lime stone and quartzite mining globally. The result shows that mining industries use financial, ecological, and societal criteria in the selection of sustainable suppliers, which are in tune with the research carried out by Dickosn (1966) and Frohling and Schultmann (2016). Therefore, the SWARA method was used for calculating the weights and corresponding ranking of the criteria along with sub-criteria were done accordingly. Then, the QFD technique was used for the design criteria. Minimizing the work-related problems in industrial, health, safety organization and ecological management is esteemed to be of greater importance. At last an integrated fuzzy method was used for order preference based on similarity with the method of the ideal solution (TOPSIS) in order to improve the quality of decision-making regarding the ranking of alternatives.

Sustainable Supply Chain Management in the Mining Industry.
Suchismita Satapathy and Hullash Chauhan (Authors)

3.1 INTRODUCTION

Sustainable supply chain management (SSCM) is the incorporation of environmental concerns into inter-organizational supply chain management exercises. Implementing SSCM in the mining region has been a slow process. Muchaendepi et al. (2019) have evaluated the challenges faced by effective functioning of SSCM in mining sector in Zimbabwe and recommended that the organizations and various stakeholders ensure that structural and organizational changes supported the improvement of the legal and regulatory framework for the environment, the lowering of costs associated with ecological products, and the distribution of vital resources for the effective implementation of a SSCMs. Hasan (2013) have investigated the kind of environmental running practices which can be undertaken with the aid of using companies in greening the supply chain and the way those practices have an effect on the environmental and operational overall performance of the company. Menon and Ravi (2021) have studied the barrier in the circumstance of the electronics production so that organization can perform better implement to SSCM. Flores (2021) have identified the criterion for the selections of sustainable provider in the supply chain of copper reduction in Chile, through the information obtained from the application of survey to executives in all mining factories in Antofagasta. Yaping and Bossman (2021) have analyzed the supply chain issues such as poor road network in mining community, variable delivery times and exchange rates, strict government laws, violent disputes over land rights, high transport costs due to bad road and rail network, unreliable delivery times (caused by delays in ports), absence of local suppliers of international standards, and political restrictions as a main challenges in the supply chain industry of Ghana. Hasan (2006) have investigated the kind of environmental organization practices that are undertaken by Australian corporations in greening the supplier chain and the way these practices have an effect on the environmental and operational performance of those companies. Karim et al. (2021) have proposed the challenges of the SSCM and their interrelations in the apparel division of Bangladesh. Pramudiawardani et al. (2020) have analyzed the academic literature that addresses the practice of SSCM in Indonesia. Loughlin et al. (2021) have investigated how company processes become bespoke sustainability process related to supply chain management through a step-by-step approach of alignment, functioning, and maintenances. Aghdaie et al. (2014) have proposed the integrated approach to supplier clustering and classification that comprises the cluster analyses as a data mining tool and the step weight assessment ratio analysis (SWARA) and VIseKriterijumskao ptimizacija i KOmpromisno Resenje (VIKOR) as the two MADM tools. Işık and Adalı (2016) have identified the multi-criteria decision-making (MCDM) methods that deal with the process of selecting an alternative from several alternatives when conflicting criteria are present. The SWARA and *operational competitiveness rating analyses* (OCRA) technique are used for solving a hotel selection problem. Suryaningrat (2016) had proposed the quality function deployment (QFD) technique that was implemented to assess consumer expectations and develop a sustainable supply chain for agro-processed cassava products in the Indonesian food market. Nayakappa et al. (2019) have used the MOORA and the SWARA MCDM approach to calculate various alternatives of smartphones presented on online shopping websites. Arudra (2015) have proposed the QFD to collect farmers' needs and designers' skills, and a survey was carried out to identify farmers and supplier' problems and needs by a group of technicians and opinions of people were compared and, finally, the opinions of the majority were efficient and fitted into the QFD approach in order to identify the main needs and problems of farmers for the design and development of machines based on their opinions at the level of agricultural production. Muda and Mat Roji (2015) have studied to obtain the feedback from employees to determine the most appropriate criterion in selecting student for internship in industry by using the QFD. Zavadskas et al. (2016) have proposed a SWARA technique with symmetric time type-2 fuzzy sets to conclude the weights of criteria based on the opinion of a group of decision-makers. Narayanan and Jinesh (2018) have studied the continuous rejection of important raw materials "was the problem face by the companies, which carried out a process of choosing suppliers with the combinations of SWARA and Topsis, the companies was able to overcome the crisis." Júnior et al. (2020) have used the QFD technique to interact with charcoal customers to become familiar and classify the quality suitable for barbecue use. Nădăban et al. (2016) have focused on an overview of the developments of Topsis Fuzzy technique that have been applied to decision-making field. Kargi (2016) has proposed the fuzzy TOPSIS technique in order to solve the companies' problem in choosing the good supplier and dealing with linguistic variables used by decision-makers. Shamsuzzoha et al. (2021)

have proposed a technique known as the fuzzy technique for order preference due to its similarities to the ideal solution (fuzzy TOPSIS) for the selection of complex projects in organizations. Ding (2011) have developed an integrated fuzzy method for order preference based on similarity with the method of the ideal solution (TOPSIS) in order to improve the quality of decision-making regarding the ranking of alternatives. Ertuğrul and Karakaşoğlu (2007) have developed an integrated fuzzy method for academic member selection problem. The evaluations of several alternatives against several subjective criteria and the weighting of all criteria are evaluated using linguistic variables, which are represented by fuzzy numbers. Javadian et al. (2009) have introduced the fuzzy technique for the fulfilment of the order rules by similarity with the ideal solution (TopSIS) base on the new concepts of the positive and negative ideal solutions. Husin et al. (2019) have applied the fuzzy TOPSIS technique to classify the risk variables according to its frequency of occurrence by the integration of time, cost, and quality criteria at the same time and by selecting the ten important variables with the order of the most important effect. Supplier's selection is a MCDM problem and strategic importance to most companies. Since conventional supplier selection method is inadequate to address the imprecise or vague nature of linguistic assessment, a new technique called the Fuzzy Technique for Order Preference for Similarity to Ideal Solution (TOPSIS). Here Fuzzy TOPSIS technique used for the supplier selection (Sevkli et al., 2010). Aliyeva (2018) have quantified the fuzzy MCDM with TOPSIS and weighted product (WP) method is used to choose the best locations of companies' resource. The best choice of location for the dam settlement can now be better approached with the concept of technique for sorting similarity to the ideal solution (TOPSIS) by the use of fuzzy technique (Adeyanju and Adedeji, 2017). Sahin et al. (2020) have focused on fuzzy TOPSIS technique to provide weights for a given criterion and ordered the alternatives to identify the optimal solution. The alternatives are ranked among the multiple providers in the context of shipping investment. Irvanizam (2018) have used the fuzzy TOPSIS technique to order the alternatives in descending order based on their relative closeness to the ideal solution. Kolios et al. (2017) have proposed a fuzzy TOPSIS technique based on MCDM problem to analyse and prioritize the most critical failure mode identified by the failure mode and effects analysis (FMEA). Erdoğan et al. (2013) have evaluated the fuzzy TOPSIS method for order preference by similarity to ideal solution technique that ranks the alternatives for evaluation of service quality in public transportation system (PTS). Ariapour et al. (2014) have developed the fuzzy TOPSIS technique to improve the quality of decision-making for species selections. Basahel and Taylan (2016) have used the TOPSIS model to calculate the performance of company and ranking them according to their safety performance. Yavuz (2016) have proposed the use of fuzzy TOPSIS technique to select the open pit truck and the optimal solution to the problem was investigated and found that fuzzy TOPSIS with a group decision is a technique that can help decision-makers to solve various decision problems in mining.

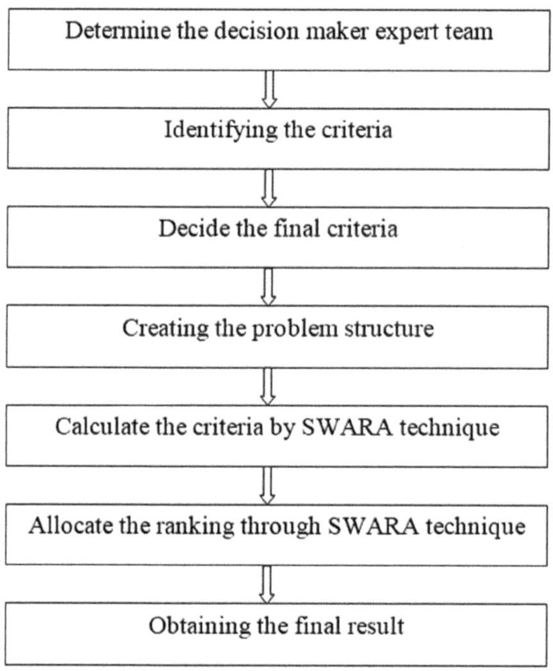

FIGURE 3.1 The SWARA technique.

The SWARA method was applied to define the weights of the criteria and sub-criteria on the basis of following steps:

Stage 1: According to the significance, the parameter is sorted, where the most important is placed first, and the least important is placed last.

Stage 2: The relative significance of average value (*sj*) is measured on the basis of the importance of the parameter *cj* compared to the criterion *cj* + 1.

Stage 3: The calculation of coefficients' (*kj*) are as below:

$$k_j = \begin{cases} 1 & j = 1, \\ s_j + 1 & j > 1 \end{cases}$$

Stage 4: The calculations of re-calculated weights (qj) are as below:

$$Q_j = \begin{cases} 1, & j = 1 \\ \dfrac{q_{j-1}}{k_j} & j > 1 \end{cases}$$

Stage 5: The calculations of comparative weights (wj) of the evaluation conditions are as below:

$$W_j = \frac{q_j}{\sum_{k=1}^{n} qk}$$

where, *n* = number of the parameter's.

3.2 METHOD

This study is based on an in-depth research to categorize the criteria for the selection for barrier of SSCM of mining sector globally. The result shows that mining industries use financial, ecological, and societal criteria for the barrier SSCM and further opinion of decision-making experts (Table 3.1). In this study, the SWARA method was used for calculating the weights and corresponding ranking of the criteria along with sub-criteria were done accordingly. Further (Table 3.2), the QFD technique was used for the design criteria to minimize the work problems in industrial health safety organization and ecological management. The fuzzy-TOPSIS Technique, that is, fuzzy-technique for order of preference by similarity to ideal solution was used to rank the most important drivers, which was a challenge for SSCM of mining sectors. Fuzzy-TOPSIS method is used due to its simplicity, rationality, comprehensibility, good computational efficiency, and ability to measure the relative performance for each alternative in a simple mathematical form.

TABLE 3.1 Decision-Making Experts.

Area	Qualifications	No. of expert
Scientist	Doctoral	1
Medical sector	MBBS	1
Environmental sector	Master degree	1
Mining sector	Master degree	1
Organizational sector	Master degree	1

TABLE 3.2 Barrier's Parameter and Sub-Parameter for Sustainable Supply Chain Management of Mining Sector.

Symbol	Parameter and sub-parameter
C_1	Capacity constraints
C_{1-1}	Poor supplier commitment
C_{1-2}	Lack of green practitioners
C_{1-3}	Poor senior management's commitment
C_{1-4}	Lack of managerial support and practical tools
C_{1-5}	Inadequate individual capacity
C_{1-6}	Customer demand for lower price
C_{1-7}	High cost associated with the implementation of SSCM
C_2	Poor legislative measures
C_{2-1}	No effective policy and legislation direction
C_{2-2}	Lack of political commitment and support
C_{2-3}	Lack of ethical and sustainability standards and appropriate regulations
C_{2-4}	Negative media attention by environmental advocacy groups
C_3	Information gap
C_{3-1}	Ineffective monitoring and control system
C_{3-2}	Lack of eco-literacy among supply chain partners
C_{3-3}	Lesser priority to R&D activities leading towards Technological innovations
C_4	Insufficient pressure from society
C_{4-1}	Lack of commitment on environmental deterioration
C_{4-2}	Lack of societal awareness regarding sustainable issues
C_{4-3}	Lack of communication and knowledge sharing

3.3 RESULTS AND DISCUSSION

As seen in Table 3.2, the most important criteria and sub-criteria for SSCM of mining sector barrier were capacity constraints, information gap, insufficient pressure from society, poor legislative measures, ineffective monitoring and control system, lack of commitment on environmental deterioration, lack of eco-literacy among supply chain partners, poor supplier commitment, lack of societal awareness regarding sustainable issues, lesser priority to R&D activities leading toward technological innovations, no effective policy and legislation direction, lack of green practitioners, lack of ethical and sustainability standards and appropriate regulations, lack of political commitment and support, Poor senior management's commitment, Lack of managerial support and practical tools, negative media attention by environmental advocacy groups, inadequate individual capacity, customer demand for lower price, high cost associated with the implementation of SSCM, and so on. Then the weights of parameter as well as sub-parameter was obtained by the use of SWARA, as illustrated in Tables 3.3–3.8. The parameter and sub-parameter based on category for SSCM of mining sector barrier were decided according to the indicate priority given by experts. Final ranking of parameter along with sub-parameter based on barrier of SSCM were found by considering their respective obtained weight values (Table 3.14)—1st, capacity constraints; 2nd, information gap; 3rd, insufficient pressure from society; 4th, poor legislative measures; 5th, ineffective monitoring and control system; 6th, lack of commitment on environmental deterioration; 7th, lack of eco-literacy among supply chain partners; 8th, lack of societal awareness regarding sustainable issues; 9th, lesser priority to R&D activities leading toward technological innovations; 10th, poor supplier commitment; 11th, lack of communication and knowledge sharing; 12th, no effective policy and legislation direction; 13th, lack of green practitioners; 14th, lack of ethical and sustainability standards and appropriate regulations; 15th, lack of political commitment and support; 16th, poor senior management's commitment; 17th, lack of managerial support and practical tools; 18th, negative media attention by environmental advocacy groups; 19th, inadequate individual capacity; 20th, customer demand for lower price; and 21st, high cost associated with the implementation of SSCM. Table 3.11 of QFD shows that the customer requirements "HOW" reduction of ecological and social conflicts in areas allocated for mining is ranked first. Potentially reduced delay in obtaining permits (environment, forestry) for mines is ranked second. Stronger monitoring systems and processes is ranked third. Improving the security of high-risk areas in terms of surroundings and social considerations is ranked fourth. Improved clarity for everyone involved about the risk levels of mining lease areas is ranked fifth. Transport and blasting of mines are done in closed chambers is ranked sixth. Simplicity and promotions of ecological, social, and governance principles (ESG) are ranked seventh. A regional plan for mineral development for selected mining areas and approach keys is ranked eight. Community ownership, profit sharing, and contribution to socio-economic development is ranked ninth. Local and cumulative impact of mining through synchronized and combined action is ranked tenth. Supply chain securities and geopolitics is ranked eleventh. Intensive use of spatial and geo-scientific information at the mine level for the assessment, planning, control, and monitoring of the mining sector is ranked twelfth. Ethical operations and responsible business practices through a proper waste disposal system is ranked thirteenth. Increase data analytics, artificial intelligence, and innovation is ranked fourteenth. Inclusion of ecological and social sensitivities in leasing decisions is ranked fifteenth. Investments in mining operations should increase is ranked sixteenth. Opportunities for grouping together small operators to become more competitive and compliant is ranked seventeenth. An outreach process that provides interested parties with relevant and timely information and enables questions to be asked in engagement forums is ranked eighteenth. The Fuzzy-TOPSIS technique to order the alternatives in descending order based on their relative closeness to the ideal solution of the final ranking of alternatives based on the closeness coefficient value indicate that "Quartzite" was ranked first, "Lime stone was ranked second," "Coal" was ranked third, "Iron" was ranked fourth, "Copper" was ranked fifth, "Gold" was ranked sixth, and "Bauxite" was ranked seventh. On basis of literature and view of experts, drivers for SSCM of mining sector were considered as the fourteen criterion as shown in (Table 3.12) correspondingly, the probable solution for the abolition of such problem as seven alternatives were measured (Table 3.13).

The overall ranking of all four parameters considered for SSCM of mining sector are—1st, capacity constraints; 2nd, information gap; 3rd, insufficient pressure from society; 4th, poor legislative measures.

TABLE 3.3 Final Weights of Parameter for Sustainable Supply Chain Management of Mining Sector.

Parameter	Relative significance of average values (s_j)	Coefficients ($k_j = s_j + 1$)	Re-calculated weights $\left(w_j = \frac{x_j - 1}{k_j}\right)$	Weights $\left(q = \frac{x_{j-1}}{\sum w_j}\right)$
C_1		1	1	0.300
C_3	0.13	1.13	0.884	0.265
C_4	0.14	1.14	0.775	0.232
C_2	0.15	1.15	0.673	0.201

TABLE 3.4 Total Weights of Categories on Capacity Constraints.

Sub parameter	Relative significance of average values (s_j)	Coefficients ($k_j = s_j + 1$)	Re-calculated weights $\left(w_j = \frac{x_j - 1}{k_j}\right)$	Weights $\left(q = \frac{x_{j-1}}{\sum w_j}\right)$	Last weights
C_{1-1}		1	1	0.216	0.064
C_{1-2}	0.16	1.16	0.862	0.186	0.055
C_{1-3}	0.19	1.19	0.724	0.156	0.046
C_{1-4}	0.20	1.20	0.603	0.130	0.039
C_{1-5}	0.13	1.13	0.533	0.115	0.034
C_{1-6}	0.12	1.12	0.475	0.102	0.030
C_{1-7}	0.11	1.11	0.427	0.092	0.027

Table 3.4 illustrates the final weights of all sub-parameter bases for sustainable supply chain management of mining sector criteria in context of the category. The final weights are based on the (C1), that is, final weight of $C_{1-1} = 0.300 \times 0.216 = 0.064$, and so on. The similar process was followed for the successive tabular calculation.

3.4 QFD ANALYSIS

QFD is a product growth organization that "organizes" the opinion of consumer through the product growth procedure. A cross-functional group implements QFD by generating a sequence of one or new matrices, the first of which is mentioned as the house of quality (HOQ). These media relate consumer wants and requirements to a wide set of product structures. A set of metrics is established to quantify how well groups of product features are summit consumer needs, and the product's design conditions are recognized and settled. QFD allows the whole product development group to select their growth activities in an orderly, efficient way. It supports them to effort together to accomplish a communal opinion of the correlation among consumer wants and product design. It provides an "audit trail" which retells people, both new and old to the project, as to why earlier results were prepared. The QFD procedure expanses the product growth group's thinking as to which activities are most serious toward producing a charming product or service besides developing communiqué among followers of the product growth group. The results are a creation that is considerately intended right from the creation, cut down of the necessity for future modifications, and decreasing growth period and cost.

TABLE 3.5 Total Weights of Categories Poor Legislative Measures.

Sub parameter	Relative significance of average values (s_j)	Coefficients ($k_j = s_j + 1$)	Re-calculated weights $\left(w_j = \frac{x_j - 1}{k_j}\right)$	Weights $\left(q = \frac{x_{j-1}}{\sum w_j}\right)$	Last weights
$C_{2\text{-}1}$		1	1	0.311	0.062
$C_{2\text{-}3}$	0.15	1.15	0.869	0.270	0.054
$C_{2\text{-}2}$	0.18	1.18	0.736	0.229	0.046
$C_{2\text{-}4}$	0.22	1.22	0.603	0.187	0.037

TABLE 3.6 Total Weights of Categories Information Gap.

Sub parameter	Relative significance of average values (s_j)	Coefficients ($k_j = s_j + 1$)	Re-calculated weights $\left(w_j = \frac{x_j - 1}{k_j}\right)$	Weights $\left(q = \frac{x_{j-1}}{\sum w_j}\right)$	Last weights
$C_{3\text{-}1}$		1	1	0.380	0.100
$C_{3\text{-}2}$	0.14	1.14	0.877	0.333	0.088
$C_{3\text{-}3}$	0.17	1.17	0.749	0.285	0.075

TABLE 3.7 Total Weights of Categories Insufficient Pressure From Society.

Sub parameter	Relative significance of average values (s_j)	Coefficients ($k_j = s_j + 1$)	Re-calculated weights $\left(w_j = \frac{x_j - 1}{k_j}\right)$	Weights $\left(q = \frac{x_{j-1}}{\sum w_j}\right)$	Last weights
$C_{4\text{-}1}$		1	1	0.392	0.090
$C_{4\text{-}2}$	0.17	1.17	0.854	0.335	0.077
$C_{4\text{-}3}$	0.23	1.23	0.694	0.272	0.063

TABLE 3.8 Overall Ranking of Parameters Along with Sub-Parameter's Based for Sustainable Supply Chain Management of Mining Sector.

Signs	Parameter along with sub parameter	Weight values	Overall position
C_1	Capacity constraints	0.300	1
C_3	Information gap	0.265	2
C_4	Insufficient pressure from society	0.232	3
C_2	Poor legislative measures	0.201	4
$C_{3\text{-}1}$	Ineffective monitoring and control system	0.100	5
$C_{4\text{-}1}$	Lack of commitment on environmental deterioration	0.09	6
$C_{3\text{-}2}$	Lack of eco-literacy among supply chain partners	0.088	7
$C_{4\text{-}2}$	Lack of societal awareness regarding sustainable issues	0.077	8
$C_{3\text{-}3}$	Lesser priority to R&D activities leading towards technological innovations	0.075	9
$C_{1\text{-}1}$	Poor supplier commitment	0.064	10
$C_{4\text{-}3}$	Lack of communication and knowledge sharing	0.063	11
$C_{2\text{-}1}$	No effective policy and legislation direction	0.062	12
$C_{1\text{-}2}$	Lack of Green Practitioners	0.055	13
$C_{2\text{-}3}$	Lack of ethical and sustainability standards and appropriate Regulations	0.054	14
$C_{2\text{-}2}$	Lack of political commitment and support	0.046	15
$C_{1\text{-}3}$	Poor senior management's commitment	0.046	15
$C_{1\text{-}4}$	Lack of managerial support and practical tools	0.039	16
$C_{2\text{-}4}$	Negative media attention by environmental advocacy groups	0.037	17
$C_{1\text{-}5}$	Inadequate individual capacity	0.034	18
$C_{1\text{-}6}$	Customer demand for lower price	0.030	19
$C_{1\text{-}7}$	High cost associated with the implementation of SSCM	0.027	20

Chan and Wu (2005) described that the HOQ covers the following phases:

1. **Consumer Necessities (*WHATs*)**
   QFD starts through the consumers (Griffin and Hauser, 1992) and afterward the collection of all consumers necessities and similar requirements are gathered into groups and inscribed into sympathy diagram
2. **Consumer Important Mass**
   In command to expose the importance mass of the consumer requirements and recognize it, the consumers are questioned to give the importance for each WHAT using grade scales.
3. **Design Necessities (*HOWs*)**
   Considering the literature review, consumers' requirements and the designs requirement (*HOWs)* are identified.
4. **Relationship Matrix among WHATs and HOWs**
   In this learning, the allocated facts and signs for QFD are 0.8 used for durable [O], 0.6 used for sensible [⊙], 0.4 used for poor [Δ], and 0.2 used for actual poor [•].
5. **Mark Value for HOWs**
   Mark value signifies "How Much" for designs necessities. At this stage, the mark value for individually designs requirement was created from literature analysis and talks by farmers.
6. **Separate Rating of Design Necessities**
   Lastly, the separate ratings of design necessities of learning carpentry workspace were designed by means of the following calculations:

$$\text{Separate Rating of Design Necessities} = \sum_{j}^{n} AijXj$$

where A_{ij} and X_j signify the comparative position of the ith features with reverence to the jth consumer want in the correlation matrix and the position of jth consumer want (consumer ratings), respectively.

n = Numbers of consumers need

Based on the SWARA result, the 17 important parameters through which mining sector may get affected are shown in Table 3.9.

To satisfy and minimize the Sustainable supplier selection, 18 design requirements can be seen in Table 3.10.

TABLE 3.9 SWARA Result on the 17 Important Sub-Parametersfor Sustainable Supply Chain Management of Mining Sector.

Sr. No	Sub-parameters
1	Poor supplier commitment
2	Lack of green practitioners
3	Poor senior management's commitment
4	Lack of managerial support and practical tools
5	Inadequate individual capacity
6	Customer demand for lower price
7	High cost associated with the implementation of SSCM
8	No effective policy and legislation direction
9	Lack of political commitment and support
10	Lack of ethical and sustainability standards and appropriate regulations
11	Negative media attention by environmental advocacy groups
12	Ineffective monitoring and control system
13	Lack of eco-literacy among supply chain partners
14	Lesser priority to R&D activities leading towards technological innovations
15	Lack of commitment on environmental deterioration
16	Lack of societal awareness regarding sustainable issues
17	Lack of communication and knowledge sharing

TABLE 3.10 Design Requirements.

Design requirements	Details
1	Supply chain securities and geopolitics
2	Increase data analytics, artificial intelligence, and innovation
3	Simplicity and promotions of ecological, social, and governance principles (ESG)
4	Investments in mining operations should increase
5	Inclusion of ecological and social sensitivities in leasing decisions
6	Community ownership, profit sharing and contribution to socio-economic development
7	Ethical operations and responsible business practices through a proper waste disposal system
8	Reduction of ecological and social conflicts in areas allocated for mining
9	More clarity for everyone involved about the risk levels of mining lease areas
10	Potentially reduced delay in obtaining permits (environment, forestry) for mines

TABLE 3.10 *(Continued)*

Design requirements	Details
11	Improves the security of high risk areas in terms of surroundings and social considerations
12	A regional plan for mineral development for selected mining areas and approach keys
13	Local and cumulative impact of mining through synchronized and combined action
14	Opportunities for grouping together small operators to become more competitive and compliant
15	An outreach process that provides interested parties with relevant and timely information and enables questions to be asked in engagement forums
16	Intensive use of spatial and geo-scientific information at the mine level for the assessment, planning, control, and monitoring of the mining sector
17	Stronger monitoring systems and processes
18	Transport and blasting of mines are done in closed chambers

The HOQ design is illustrated in Figure 3.2. The HOQ (matrix) is the recognized form of QFD. HOQ is created from these main components, as described below:

- **Consumer wants (what's):** A designed list of necessities resulting from expert's feedback.
- **Design necessities (how's):** A designed usual of applicable and computable facilities which are essential satisfying what's
- **Design matrix (left matrix):** Gives consumer's insights seen in surveys. It includes the relative status of necessities.
- **Interrelationships matrix (center matrix):** It is the interrelationship among design requirement and consumer needs. A suitable measure is applied and illustrated using signs or figure. Filling this portion of the matrix involves discussions.
- **Design correlation (top) matrix:** It used to identify where design requirement support or impede each other in the system or product design.

$$\text{Consumer rating} = Z_j + \frac{1}{n-1} (\sum B_{ij} Z_j)$$

where B_{ij} = ith initial consumers rating and Z_j indicates the relationships between needs and consumers grade, respectively.

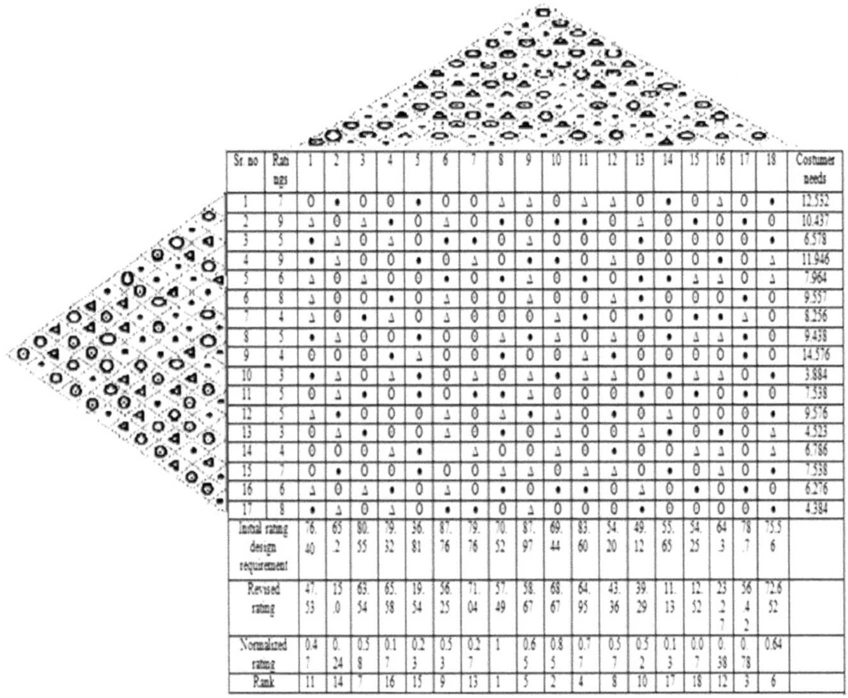

FIGURE 3.2 House of quality (HOQ).

And, n = number of consumer needs,

$$B_{ij} = [B_1ij, B_{2ij,..,} B_{nij}]$$

The separate rating for each design requirement is achieved from the center matrix using the following relation as in equation

$$\text{Individual rating of design requirements} = \sum_j^n Aij\, Xj$$

where A_{ij} and X_j signify the comparative position of the ith features with reverence to the jth consumer need in the correlation matrix and the position of jth customer needs (customer ratings), respectively.

n = Number of consumer needs

Table 3.11 shows that the customer requirements "HOW" reduction of ecological and social conflicts in areas allocated for mining is ranked first. Potentially reduced delay in obtaining permits (environment, forestry) for mines is ranked second. Stronger monitoring systems and

TABLE 3.11 Quality Function Deployment (QFD).

Sr. no	Ratings	1	2	3	4	5	6	7	8	9	10	11	12	13	14	15	16	17	18	Costumer needs
1	7	O	●	O	⊙	●	O	O	Δ	Δ	⊙	Δ	Δ	O	●	⊙	Δ	O	●	12.532
2	9	Δ	⊙	Δ	●	O	Δ	O	●	⊙	●	●	⊙	Δ	⊙	●	O	●	⊙	10.437
3	5	●	Δ	⊙	Δ	O	●	●	⊙	Δ	O	⊙	⊙	●	⊙	⊙	O	⊙	●	6.578
4	9	●	Δ	⊙	O	●	⊙	Δ	O	●	●	O	Δ	⊙	O	O	●	O	Δ	11.946
5	6	Δ	⊙	Δ	O	⊙	●	O	●	Δ	⊙	●	O	●	●	Δ	Δ	O	Δ	7.964
6	8	Δ	⊙	O	●	O	Δ	⊙	O	Δ	⊙	⊙	Δ	●	⊙	⊙	O	●	O	9.557
7	4	Δ	⊙	●	Δ	O	Δ	⊙	⊙	⊙	Δ	●	O	●	O	●	●	Δ	O	8.256
8	5	●	Δ	⊙	O	●	O	⊙	Δ	●	Δ	O	Δ	⊙	●	Δ	Δ	●	⊙	9.438
9	4	⊙	O	⊙	●	Δ	O	⊙	●	O	⊙	Δ	●	⊙	⊙	⊙	O	●	O	14.576
10	3	●	Δ	O	Δ	●	⊙	Δ	⊙	Δ	●	Δ	Δ	O	●	Δ	Δ	O	●	3.884
11	5	⊙	Δ	●	O	●	O	●	●	Δ	⊙	O	⊙	●	⊙	●	O	●	⊙	7.538
12	5	Δ	●	⊙	O	⊙	Δ	⊙	Δ	●	Δ	O	●	⊙	Δ	O	O	⊙	●	9.576
13	3	⊙	Δ	●	⊙	O	Δ	⊙	●	⊙	Δ	O	⊙	Δ	●	⊙	●	O	Δ	4.523
14	4	⊙	O	⊙	Δ	●		Δ	O	⊙	Δ	⊙	●	⊙	O	Δ	Δ	O	Δ	6.786
15	7	O	●	O	⊙	●	O	O	Δ	Δ	⊙	Δ	Δ	O	●	⊙	Δ	O	●	7.538
16	6	Δ	⊙	Δ	●	O	Δ	O	●	⊙	●	●	⊙	Δ	⊙	●	O	●	⊙	6.276
17	8	●	Δ	⊙	Δ	O	●	●	⊙	Δ	O	⊙	⊙	●	⊙	⊙	O	⊙	●	4.384
Initial rating Design Requirement		76.40	65.2	80.55	79.32	36.81	87.76	79.76	70.52	87.97	69.44	83.60	54.20	49.12	55.65	54.25	64.3	78.7	75.56	
Revised rating		47.53	15.0	63.54	65.58	19.54	56.25	71.04	57.49	58.67	68.67	64.95	43.36	39.29	11.13	12.52	23.27	56.42	72.652	
Normalized rating		0.47	0.24	0.58	0.17	0.23	0.53	0.27	1	0.65	0.85	0.77	0.57	0.52	0.13	0.07	0.38	0.78	0.64	
Rank		11	14	7	16	15	9	13	1	5	2	4	8	10	17	18	12	3	6	

processes is ranked third. Improving the security of high risk areas in terms of surroundings and social considerations is ranked fourth. More clarity for everyone involved about the risk levels of mining lease areas is ranked fifth.Transport and blasting of mines done in closed chambers are ranked sixth. Simplicity and promotions of ecological, social, and governance principles (ESG) is ranked seventh. A regional plan for mineral development for selected mining areas and approach keys is ranked eight. Community ownership, profit sharing, and contribution to socio-economic development is ranked ninth. Local and cumulative impact of mining through synchronized and combined action is ranked tenth. Supply chain securities and geopolitics is ranked eleventh. Intensive use of spatial and geo-scientific information at the mine level for the assessment, planning, control, and monitoring of the mining sector is ranked twelfth. Ethical operations and responsible business practices through a proper waste disposal system is ranked thirteenth. Increase data analytics, artificial intelligence, and innovation is ranked fourteenth. Inclusion of ecological and social sensitivities in leasing decisions is ranked fifteenth. Investments in mining operations should increase is ranked sixteenth. Opportunities for grouping together small operators to become more competitive and compliant is ranked seventeenth. An outreach process that provides interested parties with relevant and timely information and enables questions to be asked in engagement forums is ranked eighteenth.

3.5 FUZZY-TOPSIS

In this study, four experts from the medical sector, environmental sector, mining sector, and organizational sector consider, along with a scientist, were asked to give their opinion. The fuzzy-TOPSIS technique was used to rank the most important alternatives that were a challenge for SSCM of mining sector.

Stage 1: In matrix format, a representative fuzzy multiple attributes set decision-making problems was construct as:

$$\begin{array}{c} X_1 X_2 \ldots X_n \end{array}$$

$$\widehat{D} = \begin{array}{c} A_1 \\ A_2 \\ \vdots \\ A_M \end{array} \begin{bmatrix} \tilde{x}_{11} & \tilde{x}_{12} & \cdots & \tilde{x}_{1n} \\ \tilde{x}_{21} & \tilde{x}_{22} & \vdots & \tilde{x}_{2n} \\ \vdots & \vdots & \ddots & \vdots \\ \tilde{x}_{m1} & \tilde{x}_{m2} & \vdots & \tilde{x}_{mn} \end{bmatrix}$$

$$\widetilde{W} = (\widetilde{W}_1, \widetilde{W}_2, \ldots, \widetilde{W}_n)$$

Here, A_1, A_2,…, A_m denoted the feasible alternative to be select, X_1, X_2…, X_n denoted the appraisal attributes (criteria) measuring the alternative performances and corresponded to the fuzzy rating of the ith alternative A_i versus the jth attribute, X_j and $\widetilde{W}_j$ denoted the attribute weight of X_j. Here, $\bar{x}_{ij}$; $\bar{\bar{x}}_{ij}, \forall_{i,} j = 1; 2; : : : ; n$ were assessed in linguistic term that were describe by triangular fuzzy number, that is,

$$\bar{x}_{ij} = (a_{ij}, b_{ij}, c_{ij}), \widetilde{W}_j = (w_{j1}, w_{j2}, w_{j3}).$$

Stage 2: In order to evaluate the important weights of the attribute, a group of k experts was established such as:

$$E = (E_1, E_2, \ldots, E_k)$$

In addition, each group member was assigned with different voting power weights according to their professional titles as:

$$\bar{\lambda} = (\bar{\lambda}_1, \bar{\lambda}_2, \ldots.. \bar{\lambda}_k)$$

where, $\tilde{\lambda}_t$ represent the selection power weights of the t_{th} conclusion maker that was expressed by triangular fuzzy numbers.

Stage 3: The fuzzy group estimation matrix by all experts can be articulated as:

$$\begin{array}{c} \quad\quad\quad X_1 X_2 \ldots X_n \\ \tilde{w}^* = \begin{array}{c} E_1 \\ E_2 \\ \vdots \\ E_M \end{array} \begin{bmatrix} \tilde{w}^*_{11} & \tilde{w}^*_{12} & \cdots & \tilde{w}^*_{1n} \\ \tilde{w}_{21} & \tilde{w}^*_{11} & \vdots & \tilde{w}^*_{11} \\ \vdots & \vdots & \ddots & \vdots \\ \tilde{w}^*_{k1} & \tilde{w}^*_{k2} & \vdots & \tilde{w}_{Kn} \end{bmatrix} \end{array}$$

where, $\tilde{w}^*_{tj}$ indicate the fuzzy weight by the tth evaluator to the jth assessed attributes.

Stage 4: The following equation was used to integrate all the experts' opinions in order to combined the slanted judgement of k experts' to obtain the fuzzy weight ($\widetilde{W}_j$) of criteria (X_j).

$$\widetilde{W}_j = \sum_{t=1}^{k} \tilde{\lambda}_t \otimes \tilde{w}^*_{tj}, \; j = 1, 2, \ldots.n$$

Stage 5: Then, by applying the linear scale transformations method, the normalization of fuzzy decision matrix was performed as it conserves the property that the values of improved triangular fuzzy numbers remain in the range [0, 1]. Thus, the normalized fuzzy decisions matrix ($\tilde{R}$) could be recognized as:

$\tilde{R} = |\tilde{r}_{ij}|$ m x n, i = 1, 2, . . . , m; j = 1, 2, . . ., n

$\tilde{r}_{ij} = (a_{ij}/c_j^+, b_{ij}/c_j^+, c_{ij}/c_j^+), \quad j \in J^+$

$(a_j/c_{ij}, a_j/b_{ij}, a_j/c_{ij}) \quad j \in J^-$

where, $\{c_j^+ = \max_i c_{ij}, a_j = \max_i a_{ij}, J^+\}$ was associated with benefit or maximization attributes or criteria and J^- was associate with cost or minimization attributes or criteria.

Stage 6: Then by multiplying the normalized fuzzy decisions element with the aggregative fuzzy weights of each criterion, the weighted normalized fuzzy decision matrix $\tilde{v}$ can be computed as:

$\tilde{v} = |\tilde{v}_{ij}|_{m \times n}$ i = 1, 2, … , m, j = 1, 2, …, n.

where $\max_i \tilde{v}_{ij} = \tilde{w}_{ij} \otimes \tilde{r}_{ij}$, and $\tilde{v}_{ij}$, $\forall_i$ j were positive triangular fuzzy numbers.

Stage 7: Further, the fuzzy positive ideal solution (FPIS, A^+), as well as the fuzzy negative ideal solution (FNIS, A^-) were obtained as:

$A^+ = \{(\max_i \tilde{v}_{ij} | j \in J^+), \min_i \tilde{v}_{ij} | j \in J^-)\}$

$= \tilde{v}_1^+, \tilde{v}_2^+, \ldots, \tilde{v}_j^+, \ldots \tilde{v}_n^+,\}$

$A^- = \{(\min_i \tilde{v}_{ij} | j \in J^+), (\max_i \tilde{v}_{ij}) | j \in J^-)\}$

$= \{\tilde{v}_1^-, \tilde{v}_2^-, \ldots, \tilde{v}_j^-, \ldots \tilde{v}_n^-,\}$

By considering the ranges of decision elements ($\tilde{v}_{ij}$, $\forall_i$ j) belonging to the closed interval [0, 1] satisfied that:

$[\tilde{v}_I^+ = \tilde{v}_{I-}^- = (1, 1, 1)]$, and $[\tilde{v}_{j+}^- = \tilde{v}_{j-}^+ = \tilde{v}_{j-}^+ = (0; 0; 0)]$.

where, J^+ was related with benefit criteria, and J^- was related with cost criteria.

Stage 8: Then, the distances of each alternative from (A^+) and (A^-) were obtained by the "Euclidean Distance Method" as:

$$S_i^+ = \sqrt{\sum_{j=1}^{n} d^2(\tilde{v}_{ij}, \tilde{v}_j^+)}, \text{ i} = 1, 2\ldots, \text{m}$$

$$S_i^- = \sqrt{\sum_{j=1}^{n} d^2(\tilde{v}_{ij}, \tilde{v}_j^-)}, i = 1, 2\ldots, \text{m}$$

where, d ($\tilde{v}_A$, $\tilde{v}_B$) denoted the calculated distance between two triangular fuzzy number, that is, $\tilde{A}$ and $\tilde{B}$.

Stage 9: After calculating the S_i^+ and S_i^- of every alternative, a closeness coefficient (c_i^*) was considered to conclude the final place order of all alternative as:

$$c_i^* = S_i^- / (S_i^+ + S_i^-), 0 < c_i^* < 1$$

It may be noted that as (c_i^*) approaches to 1, the alternative (A_i) gets closer to (A^+) and gets farther from (A^-), respectively. Therefore, according to the corresponding closeness coefficients, the ranking order of all alternatives can be obtained.

TABLE 3.12 Selections of Criteria.

Symbol	Criteria
X1	Business comprehension
X2	Social concern to protect the natural environment
X3	Government and policy
X4	Economic welfare
X5	Competitor
X6	Workplace safety
X7	Green competence with supply chain partners
X8	Accessibility of clean technologies
X9	Maintain and initiative from various organizations
X10	Sustainable growth and development practices
X11	Complexity of the supply base
X12	Numbers of products and processes
X13	Technical improvement
X14	Number of customers

TABLE 3.13 Selections of Alternatives.

Symbol	Alternatives
A1	Lime stone
A2	Coal
A3	Bauxite
A4	Quartzite
A5	Gold
A6	Copper
A7	Iron

TABLE 3.14 Linguistic Scale with Equivalent Triangular Fuzzy Number.

Linguistic range	Symbol	Triangular fuzzy number
Very poor	VP	[0, 0, 1]
Poor	P	[0, 1, 3]
Rather poor	RP	[1, 3, 5]
Fair	F	[3, 5, 7]
Rather good	RG	[5, 7, 9]
Good	G	[7, 9, 10]
Very good	VG	[9, 9, 10]

The integrated matrix was obtaining between the criteria and alternatives, with the equivalent weights (W_i) of all the 14 criteria as shown in Table 3.15, which was follow by the formulation of normalized matrixes shown in Table 3.16, and weight normalized matrixes shown in Table 3.17.

The final ranking of alternatives (Table 3.30) based on the closeness coefficient value indicate that "quartzite" was ranked first, "limestone" was ranked second, "coal" was ranked third, "iron" was ranked fourth, "copper" was ranked fifth, "gold" was ranked sixth, and "bauxite" was ranked seventh.

3.6 CONCLUSION

A study was carried out by the use of SWARA technique to rank sustainable supplier selection. The result from overall ranking of all the ten criteria considered were found as: 1st, capacity constraints; 2nd, long working hours; 3rd, financial uncertainty; 4th, store problems; 5th, planning problems; 7th, rejection rate; 10th, organizational demands; 13th, tools and machinery; 16th, lack of informations and education; and 18th, communication between stakeholder and exposure to physical risks. These techniques are simple and most applicable on overranking process, and the second method is QFD of the customer requirements "HOW," that is, proper safety procedures used by all laborers at all times is ranked first. Proper training that must be provided is ranked second. Eco-friendly tools used is ranked third. Subsidized loan and pension scheme is ranked fourth. Proper educations are ranked fifth. Intelligence system is ranked sixth. Advanced and modern equipments is ranked seventh. Nation-wide

TABLE 3.15 Obtained Integrated Matrix.

	W_1			W_2			W_3			W_4			W_5			W_6			W_7		
	0.85	**0.64**	**1**	**0.83**	**0.67**	**1**	**0.94**	**0.78**	**1**	**0.97**	**0.65**	**1**	**0.84**	**0.54**	**1**	**0.97**	**0.65**	**1**	**0.94**	**0.75**	**1**
	X_1			X_2			X_3			X_4			X_5			X_6			X_7		
A_1	14.8000	20.2000	24.6000	2.4000	6.4000	11.8000	2.8000	6.4000	12.0000	6.8000	12.0000	17.8000	8.0000	12.8000	18.0000	18.2000	24.8000	31.2000	16.4000	23.4000	30.4000
A_2	8.4000	14.8000	22.4000	10.8000	16.6000	22.2000	16.2000	22.4000	27.8000	15.0000	20.8000	25.4000	19.0000	25.0000	29.8000	14.0000	20.6000	27.4000	12.2000	17.2000	20.8000
A_3	15.8000	22.4000	29.0000	3.4000	7.6000	13.0000	3.2000	7.4000	13.0000	2.4000	6.4000	11.8000	11.6000	17.0000	23.0000	13.2000	19.0000	25.2000	2.6000	5.4000	9.2000
A_4	23.2000	32.0000	40.8000	19.0000	24.4000	28.0000	17.0000	22.6000	26.8000	19.4000	24.6000	28.0000	9.0000	12.4000	15.6000	16.0000	20.2000	23.2000	19.6000	24.4000	28.8000
A_5	19.2000	25.8000	30.6000	6.8000	12.2000	18.2000	3.6000	8.0000	14.2000	2.8000	7.0000	12.6000	6.4000	9.8000	13.8000	9.8000	14.2000	19.0000	14.6000	19.8000	25.0000
A_6	20.2000	27.8000	35.4000	15.6000	22.2000	29.2000	8.0000	13.0000	18.6000	14.8000	21.4000	28.4000	14.6000	20.0000	25.6000	9.6000	15.0000	21.2000	16.0000	20.8000	25.0000
A_7	8.4000	14.8000	22.4000	13.4000	19.2000	25.2000	14.8000	21.0000	27.4000	17.8000	23.4000	28.4000	11.6000	17.0000	23.0000	11.6000	16.6000	21.8000	14.4000	19.0000	23.6000
Max	23.2000	32.0000	40.8000	19.0000	24.4000	29.2000	17.0000	22.6000	27.8000	19.4000	24.6000	28.4000	19.0000	25.0000	29.8000	18.2000	24.8000	31.2000	19.6000	24.4000	30.4000
Min	8.4000	14.8000	22.4000	2.4000	6.4000	11.8000	2.8000	6.4000	12.0000	2.4000	6.4000	11.8000	6.4000	9.8000	13.8000	9.6000	14.2000	19.0000	2.6000	5.4000	9.2000

TABLE 3.16 Obtained Integrated Matrix.

	W_8			W_9			W_{10}			W_{11}			W_{12}			W_{13}			W_{14}		
	0.95	0.69	1	0.78	0.55	1	0.93	0.87	1	0.89	0.76	1	0.96	0.65	1	0.88	0.63	1	0.97	0.58	1
	X_8			X_9			X_{10}			X_{11}			X_{12}			X_{13}			X_{14}		
A_1	15.0000	21.0000	27.0000	13.2000	19.0000	25.2000	16.4000	22.6000	29.0000	6.8000	12.0000	17.8000	16.4000	23.4000	30.4000	8.0000	12.8000	18.0000	2.8000	6.4000	12.0000
A_2	13.6000	18.2000	22.8000	16.0000	20.2000	23.2000	13.6000	18.2000	22.8000	15.0000	20.8000	25.4000	12.2000	17.2000	20.8000	19.0000	25.0000	29.8000	16.2000	21.8000	26.0000
A_3	13.2000	16.8000	20.4000	13.2000	16.8000	20.4000	16.2000	21.8000	26.0000	2.4000	6.4000	11.8000	2.6000	5.4000	9.2000	11.6000	17.0000	23.0000	3.2000	7.4000	13.0000
A_4	16.4000	22.6000	29.0000	16.4000	22.6000	29.0000	19.0000	25.0000	29.8000	19.4000	24.6000	28.0000	19.6000	24.4000	28.8000	9.0000	12.4000	15.6000	17.0000	22.6000	26.8000
A_5	21.0000	27.4000	33.0000	13.6000	18.2000	22.8000	11.6000	17.0000	23.0000	15.8000	22.4000	29.0000	6.8000	12.2000	18.2000	6.4000	9.8000	13.8000	2.8000	7.0000	12.6000
A_6	10.6000	16.2000	22.2000	3.6000	8.0000	14.2000	13.8000	19.4000	24.6000	4.6000	7.2000	11.4000	16.0000	20.2000	23.2000	4.0000	6.4000	10.4000	16.2000	21.8000	26.0000
A_7	10.0000	15.2000	20.8000	12.6000	17.4000	22.2000	8.8000	13.2000	18.6000	13.0000	19.0000	24.6000	13.4000	19.0000	25.0000	18.8000	25.4000	30.6000	3.2000	7.4000	13.0000
Max	21.0000	27.4000	33.0000	16.4000	22.6000	29.0000	19.0000	25.0000	29.8000	19.4000	24.6000	29.0000	19.6000	24.4000	30.4000	19.0000	25.4000	30.6000	17.0000	22.6000	26.8000
Min	10.0000	15.2000	20.4000	3.6000	8.0000	14.2000	8.8000	13.2000	18.6000	2.4000	6.4000	11.4000	2.6000	5.4000	9.2000	4.0000	6.4000	10.4000	2.8000	6.4000	12.0000

TABLE 3.17 Obtained Normalized Matrix.

	W_1			W_2			W_3			W_4			W_5			W_6			W_7		
	0.85	0.64	1	0.83	0.67	1	0.94	0.78	1	0.97	0.65	1	0.84	0.54	1	0.97	0.65	1	0.94	0.75	1
	X_1			X_2			X_3			X_4			X_5			X_6			X_7		
A_1	0.3627	0.6313	1.0603	0.0822	0.2623	0.6211	0.1007	0.2832	0.7059	0.2394	0.4878	0.9175	0.2685	0.5120	0.9474	0.5833	1.0000	1.7143	0.5395	0.9590	1.5510
A_2	0.2059	0.4625	0.9655	0.3699	0.6803	1.1684	0.5827	0.9912	1.6353	0.5282	0.8455	1.3093	0.6376	1.0000	1.5684	0.4487	0.8306	1.5055	0.4013	0.7049	1.0612
A_3	0.3873	0.7000	1.2500	0.1164	0.3115	0.6842	0.1151	0.3274	0.7647	0.0845	0.2602	0.6082	0.3893	0.6800	1.2105	0.4231	0.7661	1.3846	0.0855	0.2213	0.4694
A_4	0.5686	1.0000	1.7586	0.6507	1.0000	1.4737	0.6115	1.0000	1.5765	0.6831	1.0000	1.4433	0.3020	0.4960	0.8211	0.5128	0.8145	1.2747	0.6447	1.0000	1.4694
A_5	0.4706	0.8063	1.3190	0.2329	0.5000	0.9579	0.1295	0.3540	0.8353	0.0986	0.2846	0.6495	0.2148	0.3920	0.7263	0.3141	0.5726	1.0440	0.4803	0.8115	1.2755
A_6	0.4951	0.8688	1.5259	0.5342	0.9098	1.5368	0.2878	0.5752	1.0941	0.5211	0.8699	1.4639	0.4899	0.8000	1.3474	0.3077	0.6048	1.1648	0.5263	0.8525	1.2755
A_7	0.2059	0.4625	0.9655	0.4589	0.7869	1.3263	0.5324	0.9292	1.6118	0.6268	0.9512	1.4639	0.3893	0.6800	1.2105	0.3718	0.6694	1.1978	0.4737	0.7787	1.2041

TABLE 3.18 Obtained Normalized Matrix.

	W_8			W_9			W_{10}			W_{11}			W_{12}			W_{13}			W_{14}		
	0.95	0.69	1	0.78	0.55	1	0.93	0.87	1	0.89	0.76	1	0.96	0.65	1	0.88	0.63	1	0.97	0.58	1
	X_8			X_9			X_{10}			X_{11}			X_{12}			X_{13}			X_{14}		
A_1	0.4545	0.7664	1.2857	0.4552	0.8407	1.5366	0.5503	0.9040	1.5263	0.2345	0.4878	0.9175	0.5395	0.9590	1.5510	0.2614	0.5039	0.9474	0.1045	0.2832	0.7059
A_2	0.4121	0.6642	1.0857	0.5517	0.8938	1.4146	0.4564	0.7280	1.2000	0.5172	0.8455	1.3093	0.4013	0.7049	1.0612	0.6209	0.9843	1.5684	0.6045	0.9646	1.5294
A_3	0.4000	0.6131	0.9714	0.4552	0.7434	1.2439	0.5436	0.8720	1.3684	0.0828	0.2602	0.6082	0.0855	0.2213	0.4694	0.3791	0.6693	1.2105	0.1194	0.3274	0.7647
A_4	0.4970	0.8248	1.3810	0.5655	1.0000	1.7683	0.6376	1.0000	1.5684	0.6690	1.0000	1.4433	0.6447	1.0000	1.4694	0.2941	0.4882	0.8211	0.6343	1.0000	1.5765
A_5	0.6364	1.0000	1.5714	0.4690	0.8053	1.3902	0.3893	0.6800	1.2105	0.5448	0.9106	1.4948	0.2237	0.5000	0.9286	0.2092	0.3858	0.7263	0.1045	0.3097	0.7412
A_6	0.3212	0.5912	1.0571	0.1241	0.3540	0.8659	0.4631	0.7760	1.2947	0.1586	0.2927	0.5876	0.5263	0.8279	1.1837	0.1307	0.2520	0.5474	0.6045	0.9646	1.5294
A_7	0.3030	0.5547	0.9905	0.4345	0.7699	1.3537	0.2953	0.5280	0.9789	0.4483	0.7724	1.2680	0.4408	0.7787	1.2755	0.6144	1.0000	1.6105	0.1194	0.3274	0.7647

TABLE 3.19 Obtained Weighted Normalized Matrix.

	X1			X2			X3			X4			X5			X6			X7		
A_1	0.3083	0.4040	1.0603	0.0682	0.1757	0.6211	0.0947	0.2209	0.7059	0.2323	0.3171	0.9175	0.2255	0.2765	0.9474	0.5658	0.6500	1.7143	0.5071	0.7193	1.5510
A_2	0.1750	0.2960	0.9655	0.3070	0.4558	1.1684	0.5478	0.7731	1.6353	0.5123	0.5496	1.3093	0.5356	0.5400	1.5684	0.4353	0.5399	1.5055	0.3772	0.5287	1.0612
A_3	0.3292	0.4480	1.2500	0.0966	0.2087	0.6842	0.1082	0.2554	0.7647	0.0820	0.1691	0.6082	0.3270	0.3672	1.2105	0.4104	0.4980	1.3846	0.0804	0.1660	0.4694
A_4	0.4833	0.6400	1.7586	0.5401	0.6700	1.4737	0.5748	0.7800	1.5765	0.6626	0.6500	1.4433	0.2537	0.2678	0.8211	0.4974	0.5294	1.2747	0.6061	0.7500	1.4694
A_5	0.4000	0.5160	1.3190	0.1933	0.3350	0.9579	0.1217	0.2761	0.8353	0.0956	0.1850	0.6495	0.1804	0.2117	0.7263	0.3047	0.3722	1.0440	0.4514	0.6086	1.2755
A_6	0.4208	0.5560	1.5259	0.4434	0.6096	1.5368	0.2705	0.4487	1.0941	0.5055	0.5654	1.4639	0.4115	0.4320	1.3474	0.2985	0.3931	1.1648	0.4947	0.6393	1.2755
A_7	0.1750	0.2960	0.9655	0.3809	0.5272	1.3263	0.5004	0.7248	1.6118	0.6080	0.6183	1.4639	0.3270	0.3672	1.2105	0.3606	0.4351	1.1978	0.4453	0.5840	1.2041

TABLE 3.20 Obtained Weighted Normalized Matrix.

	X_8			X_9			X_{10}			X_{11}			X_{12}			X_{13}			X_{14}		
A_1	0.4318	0.5288	1.2857	0.3550	0.4624	1.5366	0.5118	0.7865	1.5263	0.2087	0.3707	0.9175	0.5179	0.6234	1.5510	0.2301	0.3175	0.9474	0.1013	0.1642	0.7059
A_2	0.3915	0.4583	1.0857	0.4303	0.4916	1.4146	0.4244	0.6334	1.2000	0.4603	0.6426	1.3093	0.3853	0.4582	1.0612	0.5464	0.6201	1.5684	0.5863	0.5595	1.5294
A_3	0.3800	0.4231	0.9714	0.3550	0.4088	1.2439	0.5056	0.7586	1.3684	0.0737	0.1977	0.6082	0.0821	0.1439	0.4694	0.3336	0.4217	1.2105	0.1158	0.1899	0.7647
A_4	0.4721	0.5691	1.3810	0.4411	0.5500	1.7683	0.5930	0.8700	1.5684	0.5954	0.7600	1.4433	0.6189	0.6500	1.4694	0.2588	0.3076	0.8211	0.6153	0.5800	1.5765
A_5	0.6045	0.6900	1.5714	0.3658	0.4429	1.3902	0.3620	0.5916	1.2105	0.4849	0.6920	1.4948	0.2147	0.3250	0.9286	0.1841	0.2431	0.7263	0.1013	0.1796	0.7412
A_6	0.3052	0.4080	1.0571	0.0968	0.1947	0.8659	0.4307	0.6751	1.2947	0.1412	0.2224	0.5876	0.5053	0.5381	1.1837	0.1150	0.1587	0.5474	0.5863	0.5595	1.5294
A_7	0.2879	0.3828	0.9905	0.3389	0.4235	1.3537	0.2746	0.4594	0.9789	0.3990	0.5870	1.2680	0.4232	0.5061	1.2755	0.5407	0.6300	1.6105	0.1158	0.1899	0.7647

TABLE 3.21 Calculated M (v_{ij}) Values.

	X_1	X_2	X_3	X_4	X_5	X_6	X_7
A_1	0.5909	0.2883	0.3405	0.4890	0.4831	0.9767	0.9258
A_2	0.4788	0.6437	0.9854	0.7904	0.8813	0.8269	0.6557
A_3	0.6757	0.3298	0.3761	0.2864	0.6349	0.7643	0.2386
A_4	0.9607	0.8946	0.9771	0.9186	0.4475	0.7672	0.9418
A_5	0.7450	0.4954	0.4110	0.3100	0.3728	0.5736	0.7785
A_6	0.8342	0.8633	0.6044	0.8450	0.7303	0.6188	0.8032
A_7	0.4788	0.7448	0.9457	0.8967	0.6349	0.6645	0.7445
Max	0.9607	0.8946	0.9854	0.9186	0.8813	0.9767	0.9418
Min	0.47884	0.28834	0.34048	0.2864	0.3728	0.5736	0.23859

TABLE 3.22 Calculated M (v_{ij}) Values.

	X_8	X_9	X_{10}	X_{11}	X_{12}	X_{13}	X_{14}
A_1	0.7488	0.7847	0.9415	0.4990	0.8974	0.4983	0.3238
A_2	0.6452	0.7789	0.7526	0.8041	0.6349	0.9116	0.8917
A_3	0.5915	0.6693	0.8775	0.2932	0.2318	0.6553	0.3568
A_4	0.8074	0.9198	1.0105	0.9329	0.9128	0.4625	0.9239
A_5	0.9553	0.7330	0.7214	0.8906	0.4894	0.3845	0.3407
A_6	0.5901	0.3858	0.8002	0.3171	0.7424	0.2737	0.8917
A_7	0.5537	0.7053	0.5710	0.7513	0.7349	0.9271	0.3568
Max	0.9553	0.9198	1.0105	0.9329	0.9128	0.9271	0.9239
Min	0.55371	0.38579	0.57098	0.29321	0.23178	0.27371	0.32382

TABLE 3.23 Calculated A^+ and A^- Values.

	X_1			X_2			X_3			X_4			X_5			X_6			X_7		
A^+	0.4833	0.6400	1.7586	0.5401	0.6700	1.4737	0.5478	0.7731	1.6353	0.6626	0.6500	1.4433	0.5356	0.5400	1.5684	0.5658	0.6500	1.7143	0.6061	0.7500	1.4694
A^-	0.1750	0.2960	0.9655	0	0.1757	0.6211	0.0947	0.2209	0.7059	0.0820	0.1691	0.6082	0.1804	0.2117	0.7263	0.3047	0.3722	1.0440	0.0804	0.1660	0.4694

TABLE 3.24 Calculated A^+ and A^- Values.

	X_8			X_9			X_{10}			X_{11}			X_{12}			X_{13}			X_{14}		
A^+	0.6045	0.6900	1.5714	0.4411	0.5500	1.7683	0.5930	0.8700	1.5684	0.5954	0.7600	1.4433	0.6189	0.6500	1.4694	0.5407	0.6300	1.6105	0.6153	0.5800	1.5765
A^-	0.2879	0.3828	0.9905	0.0968	0.1947	0.8659	0.2746	0.4594	0.9789	0.0737	0.1977	0.6082	0.0821	0.1439	0.4694	0.1150	0.1587	0.5474	0.1013	0.1642	0.7059

TABLE 3.25 Calculated (d^+) Values.

	X_1	X_2	X_3	X_4	X_5	X_6	X_7
A_1	0.2903	0.8592	0.7774	0.5663	0.3902	0	0.0315
A_2	0.4164	0.2542	0	0.1344	0	0.1049	0.3271
A_3	0.2003	0.7619	0.7047	1.1274	0.2038	0.1566	1.3055
A_4	0	0	0.0065	0	0.4881	0.1454	0
A_5	0.1292	0.4450	0.6335	1.0290	0.6325	0.3675	0.1744
A_6	0.0746	0.0571	0.3726	0.0954	0.1174	0.3001	0.1418
A_7	0.4164	0.1537	0.0434	0.0381	0.2038	0.2538	0.2173

TABLE 3.26 Calculated (d^+) Values.

	X_8	X_9	X_{10}	X_{11}	X_{12}	X_{13}	X_{14}
A_1	0.1913	0.0741	0.0821	0.5472	0.0278	0.4345	0.8211
A_2	0.3250	0.0566	0.2805	0.1412	0.3025	0.0096	0.0220
A_3	0.4212	0.1495	0.1256	0.9776	1.4194	0.2372	0.7231
A_4	0.1347	0	0	−4	0	0.5349	0
A_5	0	0.1014	0.3107	0.0703	0.5121	0.6758	0.7608
A_6	0.3839	0.4555	0.2173	1.0146	0.1654	0.9860	0.0220
A_7	0.4432	0.1218	0.5155	0.2046	0.1797	0	0.7231

TABLE 3.27 Calculated (d^-) Values.

	X_1	X_2	X_3	X_4	X_5	X_6	X_7
A_1	0.1411	0	0	0.2824	0.1146	0.3675	1.0732
A_2	0	0.4714	0.7774	0.7987	0.6325	0.2160	0.7974
A_3	0.1928	0.0591	0.0546	0	0.2803	0.1657	0
A_4	0.4164	0.8592	0.8101	1.1274	0.1062	0.2234	1.3055
A_5	0.2801	0.2713	0.0864	0.0300	0	0	0.9610
A_6	0	1	0	1	0	0	1
A_7	0	1	1	1	0	0	1

TABLE 3.28 Calculated (d^-) Values.

	X_8	X_9	X_{10}	X_{11}	X_{12}	X_{13}	X_{14}
A_1	0.2073	0.3439	0.4119	0.3022	1.1125	0.3335	0
A_2	0.1120	0.4054	0.2388	0.7505	0.7889	0.9979	0.7678
A_3	0	0	0.3873	0	0	0.5515	0.0417
A_4	0.2644	0.4555	0.5155	0.9776	1.4194	0.3403	0.8211
A_5	0.4432	0.3317	0	0.8003	0.4157	0	0.0248
A_6	0	0	0	0	1	0	0.7678
A_7	0	0	0	1	1	1	0.0417

TABLE 3.29 Calculations of S_i^+ and S_i^- Value.

S_i^+	S_1^+	2.0555	S_i^-	S_1^-	4.6899
	S_2^+	2.4661		S_2^-	7.7547
	S_3^+	3.8804		S_3^-	2.0904
	S_4^+	1.0721		S_4^-	9.6421
	S_5^+	2.4628		S_5^-	4.0094
	S_6^+	4.4038		S_6^-	5.9019
	S_7^+	3.5143		S_7^-	6.4813

TABLE 3.30 Ranking of Alternatives.

Alternatives	Closeness-coefficient value	Ranking
A1	1.0000	2
A2	0.9000	3
A3	0.1971	7
A4	1.4178	1
A5	0.4070	6
A6	0.5727	5
A7	0.6484	4

TABLE 3.31 Final Ranked.

Final ranked	Alternatives
1st	Quartzite
2nd	Lime stone
3rd	Coal
4th	Iron
5th	Copper
6th	Gold
7th	Bauxite

is ranked eight. Using financial reserves is ranked ninth. Legal framework to encourage or require planning at the county or municipal level is ranked tenth. Safety accessories that can save laborers from injury and protect from climate is ranked eleventh. Fuzzy-TOPSIS technique to order the alternatives in descending order base on their relative closeness to the ideal solution the final ranking of alternatives based on the closeness coefficient value indicate that "quartzite" was ranked first, "limestone" was ranked second, "coal" was ranked third, "iron" was ranked fourth, "copper" was ranked fifth, "gold" was ranked sixth, and "Bauxite" was ranked seventh.

KEYWORDS

- **TOPSIS**
- **SSCM**
- **QFD**
- **SWARA**
- **minimizing**
- **technique**

REFERENCES

Adeyanju, O. I.; Adedeji, A. A. Application of Hybrid Fuzzy-Topsis for Decision Making in Dam Site Selection. *SSRG Int. J. Civil Eng.* **2017,** *4* (7), 2348–8352.

Aghdaie, M. H.; Zolfani, S. H.; Zavadskas, E. K. Synergies of Data Mining and Multiple Attribute Decision Making. *Procedia Soc. Behav. Sci.* **2014,** *110*, 767–776.

Aliyeva, K. Facility Location Problem by Using Fuzzy Topsis Method. *Chem. Technol. Control Manag.* **2018,** *2018* (3), 55–59.

Ariapour, A. L. I.; Veisanloo, F.; Asgari, M. An Application of Fuzzy TOPSIS Method for Plant Selection in Rangeland Improvement (Case Study: Boroujerd Rangeland, Lorestan Province, Iran), 2014.

Arudra, A. K. Design and Selection of Agricultural Machinery Using a Quality Function Deployment Technique. *Agric. Eng. Int. CIGR J.* **2015,** *17* (4), 91–99.

Basahel, A.; Taylan, O. Using fuzzy AHP and Fuzzy TOPSIS Approaches for Assessing Safety Conditions at Worksites in Construction Industry. *Int. J. Saf. Secur. Eng.* **2016,** *6* (4), 728–745.

Dickson, G. An Analysis of Vendor Selection Systems and Decisions. *J. Purch. Supply Manag.* 1966, *2* (1), 5–17. http://doi.org/10.1111/j.1745493X. 1966. tb00818.x.
Ding, J. F. An Integrated Fuzzy TOPSIS Method for Ranking Alternatives and its Application. *J. Mar. Sci. Technol.* **2011,** *19* (4), 2.
Erdoğan, M.; Bilişik, Ö. N.; Kaya, İ.; Baraçh, H. A Customer Satisfaction Model Based on Fuzzy TOPSIS and SERVQUAL Methods. *Lect. Notes Manag. Sci.* **2013,** *5* (1), 74–83.
Ertuğrul, İ.; Karakaşoğlu, N. Fuzzy TOPSIS Method for Academic Member Selection in Engineering Faculty. In *Innovations in E-learning, Instruction Technology, Assessment, and Engineering Education*; Springer, Dordrecht, 2007; pp 151–156.
Gahona-Flores, O. Selection Criteria for Sustainable Suppliers in the Supply Chain of Copper Mining in Chile. *Ingeniería e Investigación* **2021,** *41* (2).
Hasan, M. Green Supply Chain Management Practices and Operational Performance in Australian Companies; Management, Pragmatism, Philosophy, Priorities, 2006.
Hasan, M. Sustainable Supply Chain Management Practices and Operational Performance. *Am. J. Ind. Bus. Manag.* **2013,** *3*, 42–48 http://dx.doi.org/10.4236/ajibm.2013.31006.
Husin, S.; Fachrurrazi, F.; Rizalihadi, M.; Mubarak, M. Implementing Fuzzy TOPSIS on Project Risk Variable Ranking. *Adv. Civil Eng.* **2019,** *2019*.
Irvanizam, I. In *Application of the Fuzzy Topsis Multi-Attribute Decision Making Method to Determine Scholarship Recipients*, Journal of Physics: Conference Series; IOP Publishing, 2018; vol *978* (1), p 012056.
Işık, A. T.; Adalı, E. A. A New Integrated Decision Making Approach Based on SWARA and OCRA Methods for the Hotel Selection Problem. *Int. J. Adv. Oper. Manag.* **2016,** *8* (2), 140–151.
Javadian, N.; Kazemi, M.; Khaksar-Haghani, F.; Amiri-Aref, M.; Kia, R. In *A General Fuzzy TOPSIS Based on New Fuzzy Positive and Negative Ideal Solution*, 2009 IEEE International Conference on Industrial Engineering and Engineering Management; IEEE, 2009; pp 2271–2274.
Júnior, A. F. D.; Andrade, C. R.; Milan, M.; Brito, J. O.; de Andrade, A. M.; de Souza, N. D. Quality Function Deployment (QFD) Reveals Appropriate Quality of Charcoal Used in Barbecues. *Scientia Agricola* **2020,** *77* (6), e20190021–e20190021.
Kargi, V. S. A. Supplier Selection for a Textile Company Using the Fuzzy TOPSIS Method. *Yönetim ve Ekonomi* **2016,** *23* (3), 789–803.
Karim R.; Ahammed T.; Hossen S.; Sakib H.; Mohaimin S. M. Challenges To Sustainable Supply Chain Management And Their Interrelation In The Apparel Sector Of Bangladesh: A Dematel Approach. *Eur. Sci. J.* **2021,** *17* (15), 409. https://doi.org/10.19044/esj.2021.v17n15p409.
Kolios, A. J., Umofia, A.; Shafiee, M. Failure Mode and Effects Analysis Using a Fuzzy-TOPSIS Method: A Case Study of Subsea Control Module. *Int. J. Multicriteria Decis. Mak.* **2017,** *7* (1), 29–53.
Mc Loughlin, K.; Lewis, K.; Lascelles, D.; Nudurupati, S. Sustainability in Supply Chains: Reappraising Business Process Management. In *Production Planning & Control*, 2021; pp 1–34.
Menon, R. R.; Ravi, V. An Analysis of Barriers Affecting Implementation of Sustainable Supply Chain Management in Electronics Industry: A Grey-DEMATEL Approach. *J. Model. Manag.* **2021,** *17* (4).

Muchaendepi, W.; Mbowa, C.; Kanyepe, J.; Mutingi, M. Challenges Faced by the Mining Sector in Implementing Sustainable Supply Chain Management in Zimbabwe. *Procedia Manuf.* **2019,** *33*, 493–500.

Muda, N.; Mat Roji, N. S. A Quality Function Deployment (QFD) Approach in Determining the Employer's Selection Criteria. *J. Ind. Eng.* **2015,** *2015*.

Nădăban, S.; Dzitac, S.; Dzitac, I. Fuzzy TOPSIS: A General View. *Procedia Comput. Sci.* **2016,** *91*, 823–831.

Narayanan, A. K.; Jinesh, N. Application of SWARA and TOPSIS Methods for Supplier Selection in a Casting Unit. *Int. J. Eng. Res. Technol.* **2018,** *7* (5), 456–458.

Nayakappa, P. A.; Geeta, A. P.; Gaurish, A. W. MOORA and SWARA Collective MCDM Approach for Smartphone Selection. *Res. Rev. Int. J. Multidiscip.* **2019,** *4* (4), 605–610.

Pramudiawardani, S.; Abidin, Z.; Hidayanto, A. N.; Budi, N. F. A.; Fitriani, W. R. Sustainability Supply Chain Management in Indonesia: A Systematic, 2020.

Sahin, B.; Yip, T. L.; Tseng, P. H.; Kabak, M.; Soylu, A. An Application of a Fuzzy TOPSIS Multi-Criteria Decision Analysis Algorithm for Dry Bulk Carrier Selection. *Information* **2020,** *11* (5), 251.

Sevkli, M.; Zaim, S.; Turkyilmaz, A.; Satir, M. In *An Application of Fuzzy Topsis Method for Supplier Selection*, International Conference on Fuzzy Systems; IEEE, 2010; pp 1–7.

Shamsuzzoha, A.; Piya, S.; Shamsuzzaman, M. Application of Fuzzy TOPSIS Framework for Selecting Complex Project in a Case Company. *J. Glob. Oper. Strateg. Sourc.* **2021**.

Suryaningrat, I. B. Implementation of QFD in Food Supply Chain Management: A Case of Processed Cassava Product in Indonesia. *Int. J. Adv. Sci. Eng. Inf. Technol.* **2016,** *6* (3).

Yaping, Q.; Bossman, M. (Logistics and Supply Chain Management Efficiency Strategy for Ghana's Mining Industry. *Eur. J. Bus. Manag. Res.* **2021,** *6* (2), 193–202.

Yavuz, M. In *Equipment Selection by Using Fuzzy TOPSIS Method*, IOP Conference Series: Earth and Environmental Science; IOP Publishing, 2016; vol *44* (4), p 042040.

Zavadskas, E. K.; Mardani, A.; Turskis, Z.; Jusoh, A.; Nor, K. M. Development of TOPSIS Method to Solve Complicated Decision-Making Problems—An Overview on Developments from 2000 to 2015. *Int. J. Inf. Technol. Decis. Mak.* **2016,** *15* (03), 645–682.

Zimmer, K.; Frohling, M.; Schultmann, F. Sustainable Supplier Management Review of Models Supporting Sustainable Supplier Selection, Monitoring and Development. *Int. J. Prod. Res.* **2016,** *54* (5), 141242. 10.1080/00207543.2015.1079340.

CHAPTER 4

Prediction of Sustainable Supply Chain Management by Soft Computing Methods

ABSTRACT

Sustainability issues have an impact on the environment, society, and the economy. The significance of these three aspects of sustainability must be considered together. Some issues in these three pillars get more attention from companies as they are more interested in identifying issues that may affect actions related to their employees. Mining activities cause a negative environmental impact and social conflict, but they also provide economic profit for the community and secure the minerals needed for low-carbon technology. Mining in these circumstances has drawn special attention because more than any other industrial activity tends to leave a strong negative impact on the environment and society on minerals and mineral production. Adaptive network-based fuzzy inference system (ANFIS) was implemented in the context of adaptive networks. The proposed ANFIS may build an input–output mapping based on both human knowledge (in the form of if then fuzzy rules) and anticipated input-output data pairs using a hybrid learning method. The ANFIS design for modeling nonlinear functions that identify nonlinear components in a control system and predict a chaotic time series, all remarkable results. The Firefly algorithm (FA) should be further hybridized with other algorithms for multitarget, multilevel optimization problems, dynamic problems, constraint management, convergence studies, and mining sustainability forecasts.

Sustainable Supply Chain Management in the Mining Industry.

Suchismita Satapathy and Hullash Chauhan (Authors)

4.1 INTRODUCTION

The sector is being forced to implement green supply chain management due to increasing demand from stakeholders, the government, and private agencies [green supply chain management (GSCM)]. To improve economic–environmental performance and assure company sustainability, industries must implement GSCM successfully. Gandhi et al. (2015) assessed the key elements that contribute to the adoption of green supply chain management. Operations in the mining and quarrying industries have significant adverse environmental consequences. Mining companies have started implementing GSCM that embraces the green information technology system (GITS) to help generate economic benefits while seeking to minimize environmental damage. Bai et al. (2017) have addressed this issue by putting in place a framework of GSCM practices and using a new decision-support method that combines grey numbers with decision-making trial and evaluation laboratory (DEMATEL) and the New Keynesian (NK) model to assess and create an implementation model. The construction industry is considered to be one of the largest waste producers in the world, and to address this problem, the concept of environmental sustainability in the construction sector can be used in different areas. The GSCM idea has been successfully implemented in other industries; however, the construction industries have its peculiarity, which requires a specific methodology to implement these concepts, which takes into account the differences between the sectors. Benachio et al. (2019) have identified the model of GSCM can be implemented in the construction industries. The impact of the coronavirus pandemic on supply chains (SC) has been examined by (Meyer et al., 2021), with a focus on the influence of the pandemic on supply chain constructs in terms of risk, resilience, and sustainability. Sustainable supply chain management (SSCM) focuses on decreasing energy and material use as well as the detrimental effects of supply chain activities. Industry must successfully use GSCM to improve environmental performance and financial performance. However, the rate of green supply chain management adoption by Indian businesses, particularly in the mining industry, remains uninspiring (Muduli et al., 2017). For GSCM to be successfully implemented, important GSCM techniques must be developed, assessed, evaluated, and chosen. Sarpong et al. (2016) have proposed an inclusive framework of the main practices and supporting practices of the GSCM that identifies the perceived impacts of the GSCM framework on sustainable corporate performance (economic, ecological, and social-triple soil) that are pertinent for the mining industry in the emerging economic. However, sustainable management practices are required to achieve long-term benefits. Sarpong et al. (2014) have introduced a broad structure for SSCM adopting in the mining industries in Ghana. The environmental problems of the extractive and mining industries in particular are widespread and harmful. An efficient way to address these harmful environmental issues is through organizational practice that involves the broader supply chain. Green supply chain practices and their role in the strategy and operation of the mining industry have not been fully addressed (Sarpong et al., 2015). Mining occupation causes a negative impact on environmental and social conflicts, but also provides economic benefits for communities and ensures the minerals needed for low-carbon technologies (Ivic et al., 2021) have analyzed the compare and critically assess the sustainability report of 10 European mining industries for the period 2016–2018 in order to resolve the drivers of adopting of sustainability practices and their contributions to the sustainable development goals (SDG). Koochaksaraei and Izadfar, (2019) have introduced an ANFIS based on a maximum power point tracking (MPPT) controller has been introduced. To transmit the maximum power to the load, the duty cycle of the two-switch fly-back inverter, which is associated between the solar panel and the load. Sharma (2012) used the adaptive network-based fuzzy inference system (ANFIS) to classify images and then compared the results withfuzzy C mean (FCM) and k-nearest neighbor (KNN) to detect the brain tumor. Walia et al. (2015) have introduced the architecture and the basic learning process behind ANFIS, which is a fuzzy inference system implemented in the context of adaptive networks. Tiwari et al. (2018) have focused on developing a data-driven adaptive neurofuzzy system for water quality index using a range of actual data obtained from eight different monitoring stations along the River Satluj in northern India. Okwu and Adetunji (2018) have developed the artificial neural network (ANN) and the ANFIS adapted to model and optimize the distribution of products in a multichip system in the envelope. Akay et al. (2012) have proposed ANFIS to simulate and analyze the mapping among the physical properties of tactile texture and people's affective response. Jang (1993) have designed the ANFIS to construct an input–output mapping based on both human knowledge and specific data. Parvar et al. (2012) have proposed the ANFIS was used to predict the overall applicability of ice cream. During data collection on ship repair yards, Fragiadakis et al. (2014) developed an ANFIS to carry out the impact of working circumstances on workplace accidents. Jovanovi et al. (2015) have studied the neurofuzzy inferences system for testing a data set model for training. Ahin and Erol (2017) have developed the ANFIS models for forecasting the attendance rates at soccer games. Sonmez et al. (2018) proposed developing an ANFIS to calculate the concentration of cadmium (Cd) in the Filyos River in Turkey. Hmouz et al. (2011) have presented the ANFIS for providing learning content adapted for mobile learners. Xing et al. (2021) have proposed a Firefly algorithm (FA) to address the high-frequency threat of modern and complex radar networks for aircraft, and examined the radar defense measures of unmanned aerial vehicles (UAV) formations with the aim of solving the allocation of radar interference resources under system countermeasures. Nguyen et al. (2018) have introduced intelligent Firefly algorithms (IFAs) to address the problem of optimal operation of thermal generation units (OOTGU) to lower the total cost of fuel used to generate electricity. The FA is a well-known swarm-based algorithm that has gained popularity in a short period of time and has a variety of applications; it is simple to understand and apply in existing studies (Khan et al., 2016). The FA is part of the swarm-based algorithm, which is known for its strong scanning, low number of parameter settings, and ease of use. However, the FA search process has some drawbacks, such as the poor accuracy of the solution. High computing time complexity and courageous oscillation (Wang et al., 2019). Verma and Mukherjee (2016) have introduced the present work that uses the FA to reduce the overloading of the transmission network in a cluster-based electricity market by reprogramming the active power of the generator. The FAs is a nature-inspired, very robust, and efficient global swarm intelligence search method that has been successfully used to solve various NP optimization problems. However, the diversification search process used by the FA can reduce the speed and rate of convergence for large-scale optimization problems. Ezugwu et al. (2020) have studied the applications of four hybrid FAs to the task of automatically clustering large, high-density, unlabeled data sets. Gebremedhen et al. (2020) have proposed the FA as an alternative method for solving topology-based optimization problems, a hybrid algorithm based on the FA which maintains the global convergence and generates topologies of the optimization criteria (OC) method. The FA is highly regarded among researchers and turns out to be an interesting technique for practitioners and researchers to solve problems from numerous research areas such as classifications, clusters, neural networks, biomedical engineering, attention medicine, and also as another research domain (Nayak et al., 2020). The FA is a new swarm intelligence introduced by Yang in 2008. It was often used to solve various optimization problems. Since many research studies have been done presenting modified versions with the intention of improving the performance of the standard.

4.1.1 MINING AND SUSTAINABLE DEVELOPMENT

There is increasing emphasis on protecting the environment from stakeholders, with social organizations, government, and the general public. As a result, building a SSCM is now part of corporate social responsibility and a challenge for businesses, including small and medium-sized businesses enterprises (Wu et al., 2018). Greening the mining supply chain is an important path that can have positive effects. The influences of stakeholders in the administration of the SSCM are growing so that researcher takes into account economic, ecological, and social risks. Rebs et al. (2018) investigated stakeholder influence and risks in SSCM as addressed in conceptual frameworks, empirical studies, and formal models in qualitative and quantitative research. Growing awareness of social, economic, and environmental issues has led companies to implement sustainable practices. In recent years, organizations have been held liable for their company activities that have an impact on the environment, society, and the economy of their own business as well as their supply chain participants. Various stakeholders are putting pressure on organizations and their supply chains to take sustainable action in order to achieve the goals of SSCM (Saeed et al., 2017). Sustainable supply chain management has increased significantly in mining, business, and science. Flores et al. (2020) have reviewed the existing literature on SSCM in emerging markets from a global perspective. Mining is important and an important source of income for the economies of many developing countries.

Environmental, social, and economic these are all factors that impact mining.

- *Factors of environmental that impact in mining*
  - Gas emission

- Air emission
- Utilization of water and pollution
- Bad effect on biodiversity
- Waste generations
- Consequential use of energy
- Renewable energy
- SDGs involvement
- Weather change

❖ *Factors of social that impact mining*
- High risk for workers
- Draw an experienced employee
- Health and safety issues
- Labor rights issues
- Conflict
- Social authorization to operate
- Social sustainability
- Impact on local people
- Community

❖ *Factors of economic that impact mining*
- Superiority reduces cost
- Liability expenditure of energy
- Process income
- Grid income
- Sales
- Resources cost
- R&D costs

4.2 METHODOLOGY

4.2.1 MENTAL STRESS PREDICTION USING ANFIS

The correlation coefficients associated with different environmental, social, and economic parameters are factors that impact mining along with the corresponding p-vales were obtained by the utilization of Pearson product moment correlation using Minitab 2017 software. From the correlation coefficients between the environmental, social, and economic, and the associated parameters for mining, it was found that they were significant at $p = 0.05$ (Tables 4.1–4.3).

TABLE 4.1 Correlation of Environmental Impact and Their Associated Parameters.

Parameter	Correlation of environmental impact
Greenhouse gas emissions (GHG)	0.579*
Air emissions	0.098
Water consumption and pollution	0.060
Negative impact on biodiversity	0.704*
Waste generation	0.514*
Significant use of energy	0.377
Renewable energy; GHG emissions	0.422
SDGs contribution	0.558*
Climate change, SDGs	0.344

*Correlation is significant at $p < 0.05$ (two tailed).

The residual plot for mining environmental impact (Figure 4.1) revealed that most of the points were clustered around the line, indicating that the error terms/residuals were approximately normal. The residuals plotted against the fitted values in the top right graph indicated that the assumption of residuals having a mean of zero was valid. The bottom left graph reemphasized the normality assumption. The bottom right graph showed a clear cyclic pattern which indicates that the residuals are dependent on a number of mining. The correlation found that GHG, negative impact on biodiversity, and SDGs contribution are more effective in mining out of all these parameters.

From the residual plot for mining of environmental impact (Figure 4.2), it was seen in the top left graph that most of the points were clustered around the line which indicated that the error terms/residuals were approximately normal. The residuals plotted against the fitted values in the top right graph indicated that the assumption of residuals having a mean zero was valid. The bottom left graph reemphasized the normality assumption. The bottom right graph showed a clear cyclic pattern which indicates that the residuals are dependent on a number of mining. The correlation found that health and safety; human and labor rights issues; conflicts; and social sustainability are more effective in mining among all the parameters.

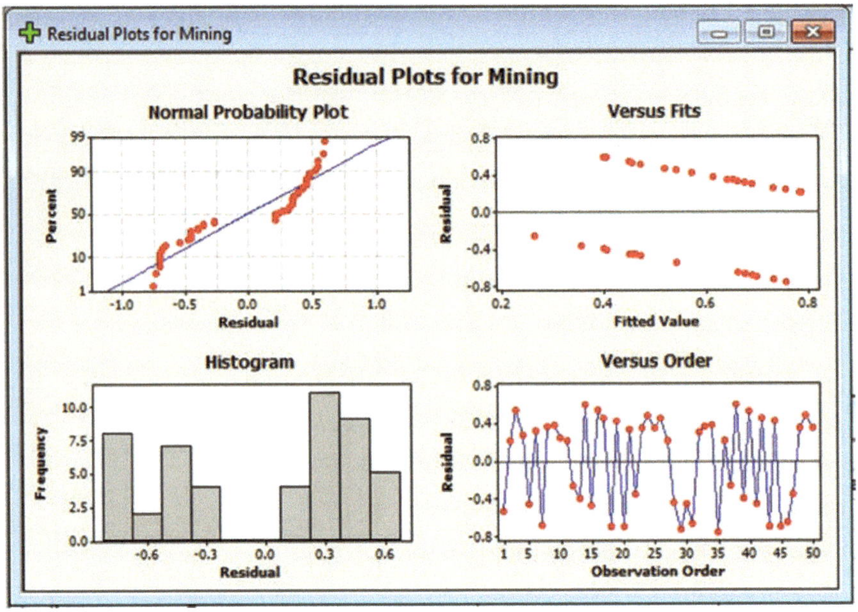

FIGURE 4.1 Residual plot for environmental impact.

TABLE 4.2 Correlation of Social Impact and Their Associated Parameters.

Parameter	Correlation of social impact
High risk for employees	0.184
Attracting and retaining high-skill workers	−0.240
Low share of female employees	0.098
Health and safety; human and labor rights issues, conflicts	0.313*
Social license to operate	0.210
Social sustainability	0.343*
Health and safety	0.039
Security/human rights and rights of indigenous peoples	0.186
Impact on local communities	0.360*

*Correlation is significant at $p < 0.05$ (two tailed).

From the residual plot for mining of environmental impact (Figure 4.3), it was seen in the top left graph that most of the points were clustered around the line which indicated that the error terms/residuals were approximately normal. The residuals plotted against the fitted values in the top right graph indicated that the assumption of residuals having a mean zero was valid. The bottom left graph reemphasized the normality assumption. The bottom right graph showed a clear cyclic pattern which indicates that the residuals are dependent on a number of mining. The correlation found that cost reductions, quality, and energy cost are more effective in mining among all the parameters.

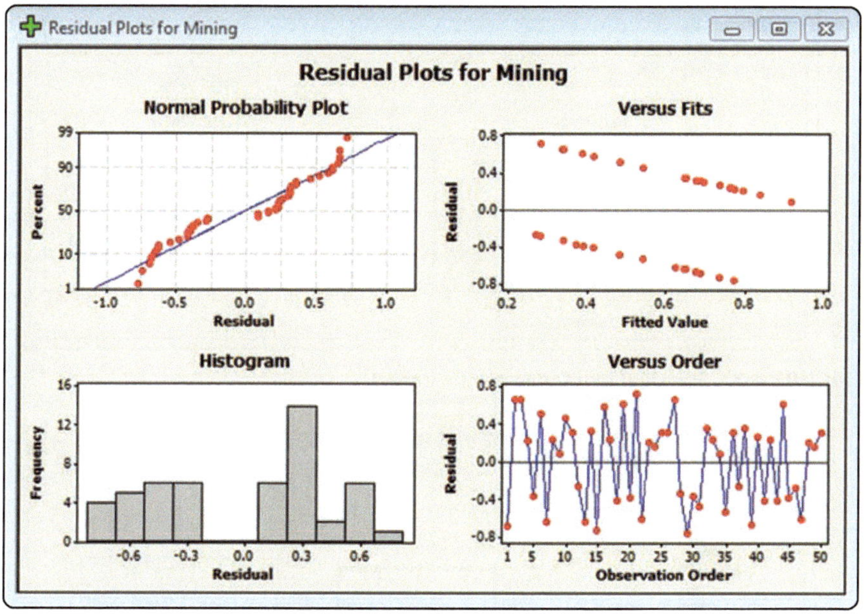

FIGURE 4.2 Residual plot for social impact.

TABLE 4.3 Correlation of Economic Impact and Their Associated Parameters.

Parameter	Correlation of economic impact
Quality	0.681*
Cost reductions	0.749*
Accountability	0.476
Energy cost	0.538*
Operation income	0.473
Net income	0.321
Sales	0.168
Raw material cost	0.321
R&D expenses	0.168

*Correlation is significant at $p < 0.05$ (two tailed).

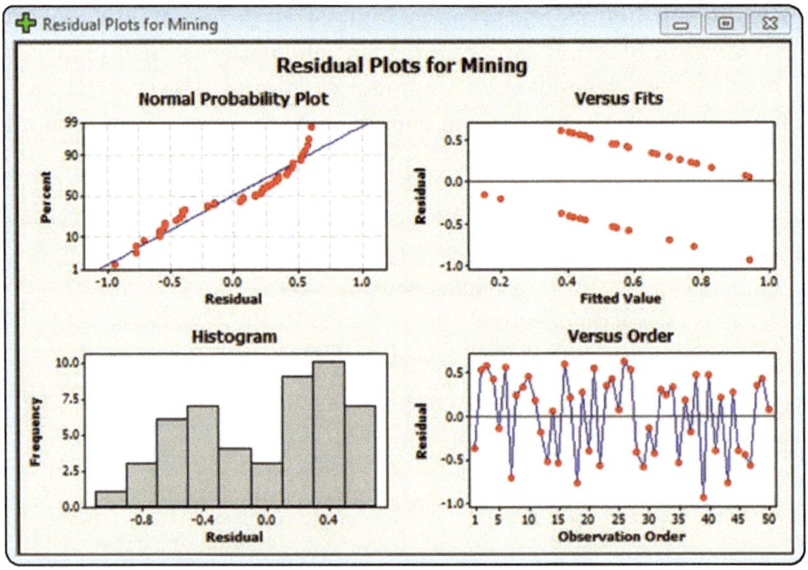

FIGURE 4.3 Residual plot for economic impact.

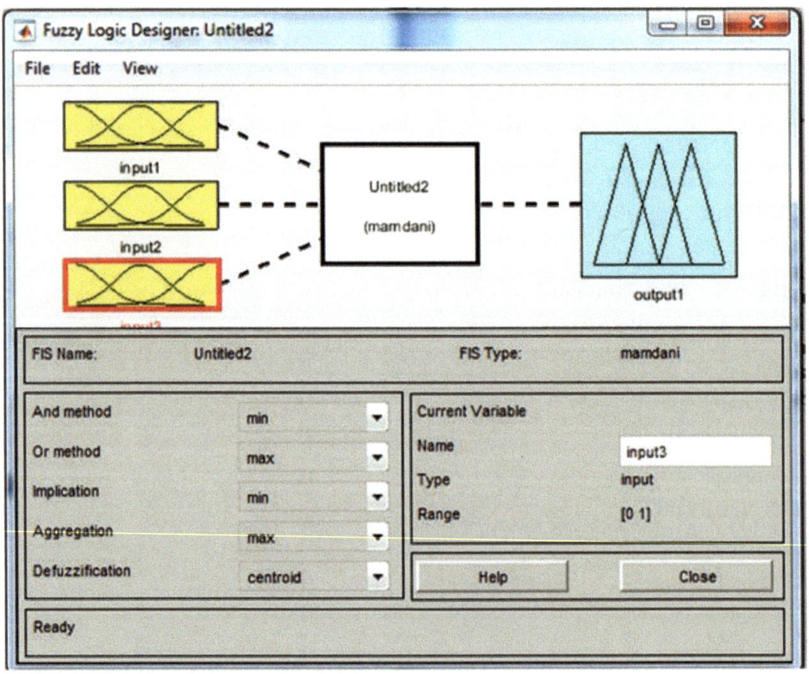

FIGURE 4.4 ANFIS-model for sustainability of mining.

For generating the ANFIS model, all parameters from environmental, social, and economic these are factors that impact mining were considered as inputs, and sustainability for mining was considered as output.

Information on ANFIS model for the sustainability prediction for mining included the following:

ANFIS information:

- Number of nodes: 1503
- Number of linear parameters: 729
- Number of nonlinear parameters: 36
- Total number of parameters: 765
- Number of training data pairs: 51
- Number of checking data pairs: 0
- Number of fuzzy rules: 729

Start training ANFIS...

1. 0.328395
2. 0.328395

Designated epoch numbers reached —> ANFIS training completed at epoch 200 with an error, respectively.

Sustainability of training data for mining:

Subsequently, after making significant adjustments in the structure of the ANFIS network to obtain the most effective model, a bell-shaped membership function with 200 epochs was selected. Sustainability prediction, errors, and ANFIS information for the ANFIS model were illustrated in Figures 4.5 and 4.6, respectively. It was observed that a better prediction with minimum errors for Sustainability for mining was achieved by ANFIS.

4.2.2 STUDIES ON FIREFLY ALGORITHM

The introduction of many metaheuristic algorithms was inspired by nature. It has solved problems without being told, but rather through experience. The primary motivation for the first metaheuristic algorithms was natural selection and survival of the fittest. Different animals communicate with one another via various communication channels. Fireflies communicate by flashing their wings. There are approximately 2000 species of Fireflies, each with its own unique lighting pattern. They typically produce a brief

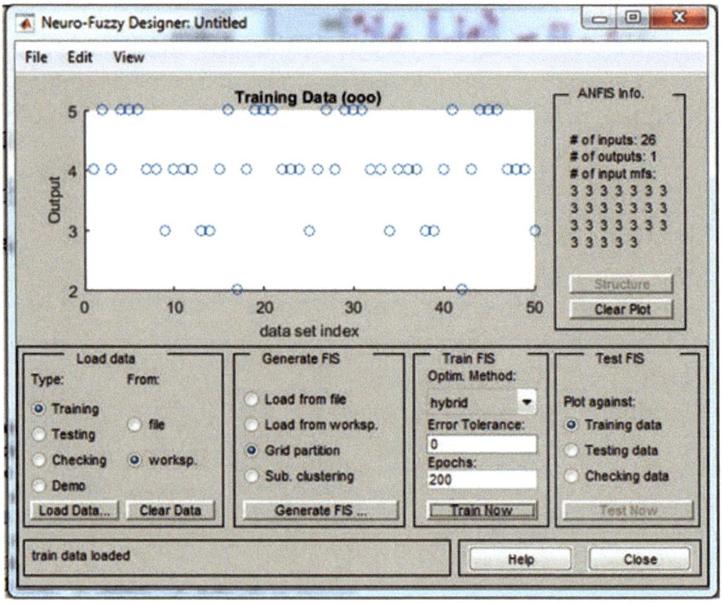

FIGURE 4.5 Sustainability of training data for mining by ANFIS model.

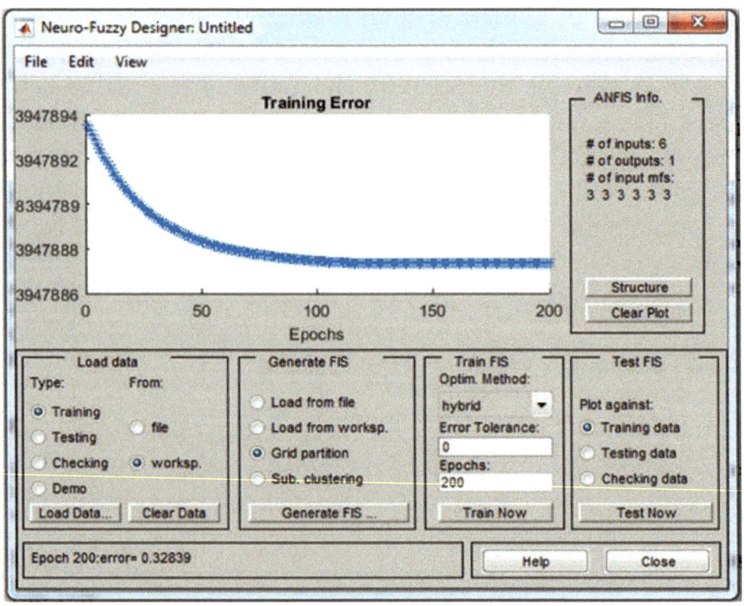

FIGURE 4.6 Sustainability predictions for mining by ANFIS-model.

flash of lightning in a predetermined pattern. Light is created through a biochemical process called bioluminescence. Flash communication is used to attract your partner and also to warn predators. A suitable partner communicates back by imitating the same pattern or responding with a specific pattern, depending on the light pattern. It should also be noted that the light intensity is a review and comparison study of the Luciérnic algorithm, and its modified versions are reduced by distance.

The standard Fireflies algorithm

The "Fireflies algorithm" is a swarm-based metaheuristic algorithm developed by Yang (2010).

The algorithm mimics how fireflies interact with their blinking lights. The algorithm assumes that all Fireflies are unisex, which means that any Firefly can be attracted to any other Firefly; a Firefly's attractiveness is directly proportional to its brightness, which is determined by the lens's function. A Firefly is attracted to a lighter color Firefly. In addition, according to the inverse square law, the brightness decreases with distance, as given in eq 4.1.

$$I \propto \frac{1}{r^2} \tag{4.1}$$

The light intensity at a distance of r from the source can be described as in eq 4.2 if the light is traveling through a medium with a light absorption coefficient.

$$I = I_0 e^{-\gamma r^2} \tag{4.2}$$

Here, I_0 is light strength at the basis. Likewise, the brightness can be given as in eq 4.3.

$$\beta = \beta_0 \, e^{-\gamma r^2} \tag{4.3}$$

A generalized brightness function for 1 is given in eq 4.4. In reality, you can use any monotonically declining function.

$$\beta = \beta_0 \, e^{-\gamma r^{\omega}} \tag{4.4}$$

Based on how well it performs in the objective function, the algorithm assigns a Firefly, a randomly produced feasible solution, and a light intensity. From this intensity, the program determines the glow worm's brightness, which is exactly proportional to its light intensity. A solution to minimization problems is given a solution with a higher light intensity and a lower function value. Each Firefly then follows the Firefly with the stronger light intensity once the solutions' intensity or brightness has been assigned. Make a local search by idly roaming around your neighborhood to find the brightest Firefly. So, for a pair of Fireflies, if Firefly *j* is brighter than Firefly *i*. Firefly *i* will migrate in the direction of Firefly *j* using the updating formula shown in eq 4.5.

$$X_i := x_i + \beta 0 e^{-\gamma rij^2}\ (x_j - x_i) + \alpha\ (€() - 0.5) \tag{4.5}$$

where β_0 is an algorithm parameter that determines how much the updating process depends on the distance between the two Fireflies, is an algorithm parameter for the step length of the random movement, and () is a random vector from a uniform distribution with values between 0 and 1. Khan et at. (2016) suggested using 0 = 1 for implementation. The second expression in equation describes the brightest Firefly, x_b. In accordance with eq 4.5 will be omitted.

$$X_b := x_b\, \alpha\ (€() - 0.5) \tag{4.6}$$

In this work, the equation (cost function) is designed through the Minitab software to get the global optimum cost values for the dependent variable sustainable. Further, the following equation also provides their corresponding local optimum cost values for independent variables: Environmental, social, and economic. The important parameters of FA and their selected values are listed in Table 4.4. These parameters help us to tune the final results of FA. Figure 4.7 shows the cost function plotting graphs for discomfort level versus the number of iterations. The developed cost function equation of the present work is given as follows:

The regression equation is as follows:

Sustainable = −4.0 − 0.87 gas emission + 0.122 air emission
+ 0.593 utilizations of water and pollution
+ 0.48 bad effect on biodiversity + 0.80 waste generations
− 0.48 consequential use of energy + 0.000 renewable energy
− 0.14 SDGs involvement − 0.52 weather change
− 0.21 high risk for workers + 0.38 attract and retain a highly quality
− 0.494 health and safety issues − 0.39 labor rights issues
+ 0.93 conflict + 0.354 social authorize to operate
+ 1.00 social sustainability −0.083 impact on local people
+ 0.31 community − 0.96 superiority + 1.04 reduces cost
− 0.73 liability − 0.604 expenditure of energy
+ 0.507 process income + 0.249 grid income

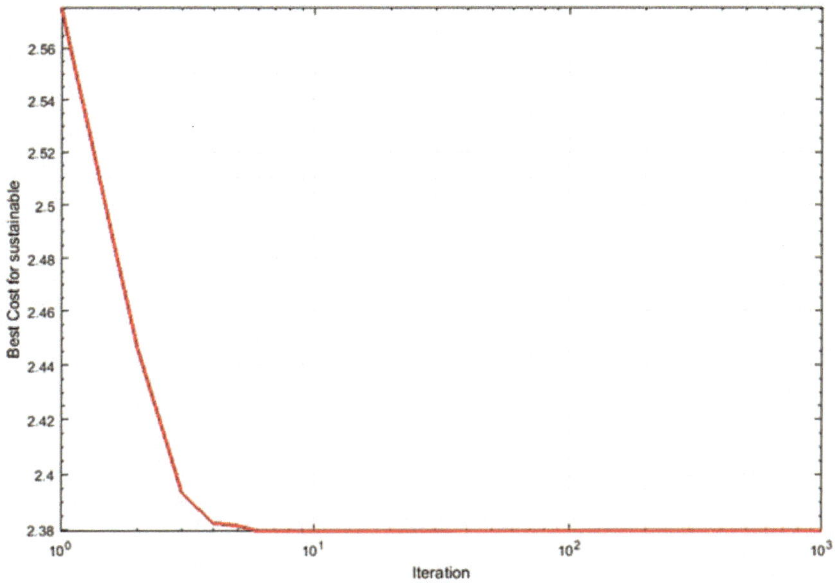

FIGURE 4.7 The cost function plotting graphs for sustainable versus the number of iterations.

TABLE 4.4 Important Parameter of FA and Their Selected Values.

Parameter	Values
Maximum number of iterations	1000
Numbers of Fireflies	135
Number of decision variables	54
Cost function	Multiobjective
Attractiveness constant, α_0	3.365934714431846e-10
Light absorption coefficient, β	1.999999932992629
Best cost	2.3794

4.3 CONCLUSION

In this study, after making significant adjustments in the structure of the ANFIS network to obtain the most effective model, a bell-shaped membership function with 200 epochs was selected. Sustainability prediction, errors, and ANFIS information for the ANFIS model were illustrated in Figures 4.5 and 4.6, respectively. It was observed that a better prediction with minimum errors for sustainability for mining was achieved by ANFIS. Further, in the FA, after the number of iterations was: 1000, the number of Fireflies was: 135, the number of decision variables was 54, the cost function was multiobjective, the attractiveness constant was 3.365934714431846e-10, light absorption coefficient was 1.999999932992629, and the best cost is 2.3794. Fireflies are used properly; the mutation operator, elitist move of the brighter Firefly method, and potential adoption of the opposition-based strategy may all lead to higher performance.

KEYWORDS

- **mining**
- **sustainability**
- **sustainable development goals (SDGS)**
- **ANFIS**
- **Firefly**

REFERENCES

Akay, D.; Chen, X.; Barnes, C.; Henson, B. ANFIS Modeling for Predicting Affective Responses to Tactile Textures. *Hum. Factors Ergon. Manuf. Service Ind.* **2012,** *22* (3), 269–281.

Al-Hmouz, A.; Shen, J.; Al-Hmouz, R.; Yan, J. Modeling and Simulation of an Adaptive Neuro-Fuzzy Inference System (ANFIS) for Mobile Learning. *IEEE Trans. Learn. Technol.* **2011,** *5* (3), 226–237.

Bahram-Parvar, M.; Salehi, F.; Razavi, S. M. Adaptive Neuro-Fuzzy Inference System (ANFIS) Simulation for Predicting Overall Acceptability of Ice Cream. *Eng. Agric. Environ. Food* **2017,** *10* (2), 79–86.

Bai, C.; Kusi-Sarpong, S.; Sarkis, J. An Implementation Path for Green Information Technology Systems in the Ghanaian Mining Industry. *J. Clean. Prod.* **2017,** *164,* 1105–1123.

Bazi, S.; Benzid, R.; Bazi, Y.; Rahhal, M. M. A. A Fast Firefly Algorithm for Function Optimization: Application to the Control of BLDC Motor. *Sensors* **2021,** *21* (16), 5267.

Benachio, G. L. F.; Freitas, M. C. D.; Tavares, S. F. In *Green Supply Chain Management in the Construction Industry: A Literature Review*, IOP Conference Series: Earth and Environmental Science; IOP Publishing, 2019; vol *225* (1), p 012011.

Ezugwu, A. E. S.; Agbaje, M. B.; Aljojo, N.; Els, R.; Chiroma, H.; Abd Elaziz, M. A Comparative Performance Study of Hybrid Firefly Algorithms for Automatic Data Clustering. *IEEE Access* **2020,** *8*, 121089–121118.

Fragiadakis, N. G.; Tsoukalas, V. D.; Papazoglou, V. J. An Adaptive Neuro-Fuzzy Inference System (anfis) Model for Assessing Occupational Risk in the Shipbuilding Industry. *Safety Sci.* **2014,** *63*, 226–235.

Gandhi, S.; Mangla, S. K.; Kumar, P.; Kumar, D. Evaluating Factors in Implementation of Successful Green Supply Chain Management Using DEMATEL: A Case Study. *Int. Strateg. Manag. Rev.* **2015,** *3* (1-2), 96–109.

Gebremedhen, H. S.; Woldemichael, D. E.; Hashim, F. M. A Firefly Algorithm Based Hybrid Method for Structural Topology Optimization. *Adv. Model. Simul. Eng. Sci.* **2020,** *7* (1), 1–20.

Ivic, A.; Saviolidis, N. M.; Johannsdottir, L. Drivers of Sustainability Practices and Contributions to Sustainable Development Evident in Sustainability Reports of European Mining Companies. *Discov. Sustain.* **2021,** *2* (1), 1–20.

Jang, J. S. ANFIS: Adaptive-Network-Based Fuzzy Inference System. *IEEE Trans. Syst. Man Cybern.* **1993,** *23* (3), 665–685.

Jovanović, B.; Grbić, T.; Bojović, N.; Kujačić, M.; Šarac, D. Application of ANFIS for the Estimation of Queuing in a Postal Network Unit: A Case Study. *Acta Polytechnica Hungarica* **2015,** *12* (7).

Khan, W. A.; Hamadneh, N. N.; Tilahun, S. L.; Ngnotchouye, J. M. A Review and Comparative Study of Firefly Algorithm and its Modified Versions. In *Optimization Algorithms-Methods and Applications*, 2016; pp 281–313.

Koochaksaraei, A. A.; Izadfar, H. In *High-Efficiency MPPT Controller Using ANFIS-Reference Model For Solar Systems*, 2019 5th Conference on Knowledge Based Engineering and Innovation (KBEI); IEEE, 2019; pp 770–775.

Kusi-Sarpong, S.; Bai, C.; Sarkis, J.; Wang, X. Green Supply Chain Practices Evaluation in the Mining Industry Using a Joint Rough Sets and Fuzzy TOPSIS Methodology. *Resour. Policy* **2015,** *46*, 86–100.

Kusi-Sarpong, S.; Sarkis, J.; Wang, X. Assessing Green Supply Chain Practices in the Ghanaian Mining Industry: A Framework and Evaluation. *Int. J. Prod. Econ.* **2016,** *181*, 325–341.

Meyer, A.; Walter, W.; Seuring, S. The Impact of the Coronavirus Crisis on Supply Chains and Their Sustainability: A Text Mining Approach. *Front. Sustain.* **2021,** *2*, 3.

Muduli, K.; Biswal, J. N.; Satapathy, S.; Barve, A.; Tripathy, S. Investigation of Influential Factors of Green Supply Chain Management in Indian Mining Industries: An Empirical Study. *Int. J. Bus. Excell.* **2017,** *12* (3), 351–375.

Nayak, J.; Naik, B.; Dinesh, P.; Vakula, K.; Dash, P. B. Firefly Algorithm in Biomedical and Health Care: Advances, Issues and Challenges. *SN Comput. Sci.* **2020,** *1* (6), 1–36.

Nguyen, T. T.; Quynh, N. V.; Van Dai, L. Improved Firefly Algorithm: A Novel Method for Optimal Operation of Thermal Generating Units. *Complexity* **2018,** *2018*.

Okwu, M. O.; Adetunji, O. A Comparative Study of Artificial Neural Network (ANN) and Adaptive Neuro-Fuzzy Inference System (ANFIS) Models in Distribution System with Nondeterministic Inputs. *Int. J. Eng. Bus. Manag.* **2018,** *10*, 1847979018768421.

Rebs, T.; Brandenburg, M.; Seuring, S.; Stohler, M. Stakeholder Influences and Risks in Sustainable Supply Chain Management: A Comparison of Qualitative and Quantitative Studies. *Bus. Res.* **2018,** *11* (2), 197–237.

Saeed, M. A.; Waseek, I.; Kersten, W. In *Literature Review of Drivers of Sustainable Supply Chain Management*, Proceedings of the Hamburg International Conference of Logistics (HICL); epubli., 2017; pp 159–184.

Şahin, M.; Erol, R. A Comparative Study of Neural Networks and ANFIS for Forecasting Attendance Rate of Soccer Games. *Math. Comput. Appl.* **2017,** *22* (4), 43.

Sánchez-Flores, R. B.; Cruz-Sotelo, S. E.; Ojeda-Benitez, S.; Ramírez-Barreto, M. Sustainable Supply Chain Management—A Literature Review on Emerging Economies. *Sustainability* **2020,** *12* (17), 6972.

Sharma, M. Artificial Neural Network Fuzzy Inference System (ANFIS) for Brain Tumor Detection, 2012. arXiv preprint arXiv:1212.0059.

Simonov Kusi-Sarpong, J. S.; Wang, X.; Leal Filho, W. Sustainable Supply Chain Management Practices in Ghana's Mining Industry, 2014.

Sonmez, A. Y.; Kale, S.; Ozdemir, R. C.; Kadak, A. E. An Adaptive Neuro-Fuzzy Inference System (ANFIS) to Predict of Cadmium (Cd) Concentrations in the Filyos River, Turkey, 2018.

Tiwari, S.; Babbar, R.; Kaur, G. Performance Evaluation of Two ANFIS Models for Predicting Water Quality Index of River Satluj (India). *Adv. Civil Eng.* **2018,** *2018*.

Verma, S.; Mukherjee, V. Firefly Algorithm for Congestion Management in Deregulated Environment. *Eng. Sci. Technol. Int. J.* **2016,** *19* (3), 1254–1265.

Walia, N.; Singh, H.; Sharma, A. ANFIS: Adaptive Neuro-Fuzzy Inference System-a Survey. *Int. J. Comput. Appl.* **2015,** *123* (13).

Wang, C.; Chu, X. An Improved Firefly Algorithm with Specific Probability and its Engineering Application. *IEEE Access* **2019,** *7*, 57424–57439.

Wu, Z.; Zhai, S.; Hong, J.; Zhang, Y.; Shi, K. Building Sustainable Supply Chains for Organizations Based on QFD: A Case Study. *Int. J. Environ. Res. Public Health* **2018,** *15* (12), 2834.

Xing, H. X.; Wu, H.; Chen, Y.; Wang, K. A Cooperative Interference Resource Allocation Method Based on Improved Firefly Algorithm. *Def. Technol.* **2021,** *17* (4), 1352–1360.

CHAPTER 5

Waste Management Strategies in the Mining Industry for Sustainability

ABSTRACT

Waste management is a strategic issue in the supply chain. The waste production is rapidly increasing, putting tremendous strain on urban development. The waste management in mining sector supply chain consists primarily of three processes: collection, transportation, and disposal. This study investigates waste definition and classification from various angles. From an objective and cumulative cost standpoint, a generic functional model for modeling waste material and flow is presented. Then, by the use of TODIM technique, as one of the most recent multi-criteria decision-making method are selected to solve the problems of the key risk-factor affecting waste management in mining sector. The results show that the purpose of this technique is quite helpful for decision making and all the identified risk-factors were ranked according to the method has succeed to get the best alternative from highest to low based on their preferences. Further based on the objective function as minimization of pollution affected by mining sector, the particle swarm optimization (PSO) was used to find the best fitness values.

5.1 INTRODUCTIONS

For several reasons, waste is a strategic problem in the supply chain. The first step in a company's cost-cutting strategy is to minimize all forms of waste in both internal and external supply chains. Second, there are more

Sustainable Supply Chain Management in the Mining Industry.
Suchismita Satapathy and Hullash Chauhan (Authors)

laws and regulations governing waste management on the national and international levels. Thirdly, the effects of goods and services on the environment are causing customers and consumers to become increasingly worried. Saif et al. (2017) have reported that the municipal solid waste (MSW) is commonly kept unintentionally stored in landfills in the past. Impacts of landfills on the economy, environment, and society have sparked interest in the sustainable use of MSW in order to add value. Xu and Yang (2022) have developed the smart grid and renewable energy systems, it is feasible to create intelligent garbage cans using IoT gadgets like smart sensors. By collecting data on how much waste is produced and collected in the first place, technology, particularly IoT devices, can be utilized to reduce the worry about waste disposal. With this data, more effective waste reduction techniques, such as separation and recycling, can be implemented in the future. Amemba et al. (2013) have examined the integration of several green supply chain components into business operations. Organizations must rearrange their supply chain operations to save the limited resources as a result of the rise in greenhouse emissions and environmental damage by businesses. For businesses in the hotel sector to guarantee that they provide customers with superior services, energy and water are their two main resource inputs. Hicks et al. (2004) have studied the waste management techniques can simultaneously lower disposal costs and produce extra value by establishing new supply chains that recycle or reuse materials. Björklund and Fors (2018) have examined a sustainable waste management strategy and how to behave in for-profit organizations while taking into account economic and ecological considerations. In addition, green supply chain management (GSCM) is used to analyze these acts from several angles in a sustainable manner. Hui (2021) has identified the goal of green supply chain management (GSCM) which aim is to decrease or eliminate waste along the supply chain, including energy, emissions, and solid wastes. GSCM has emerged as a significant new invention that aids businesses in creating "win-win" strategies that achieve profit and market share goals while reducing environmental risks and impacts and increasing ecological efficiency. Nyarko et al. (2020) have used our home-to-home waste collection model to analyzed the level of application of GSCM in Ghana's waste management value chain. Examining the social, environmental and economic implications of waste Eco-Care and other solid waste disposal companies that have incorporated a circular economy (at least ultimately a collection and recycling channel) into their services is a green dictation. It shows that it is setting the pace of understanding, recruitment, and imitation. Green supply chain management in a national waste management. Xin et al. (2021) have introduced the impacts of safe transportations and efficient management of hazardous waste on public health and the environment, urban hazardous waste reverse logistics of municipal hazardous waste (RLMHW). Mandičák et al. (2021) have examined the information and communication technologies might reduce waste production costs and contribute to a sustainable supply chain. The basic premise of the research is that information and communication technology will have a positive impact on the Slovak construction industry's ability to manage waste costs and sustainably. Yang et al. (2021) have analyzed the collection, transportation, and disposal which make up the bulk of the municipal solid waste (MSW) supply chain in China. In accordance with the classification, the garbage is sorted at the collection and disposed of. At the transportation stage, it is mixed, though. The individually collected garbage is mixed together during mixed transportation, which has a significant impact on how the waste is disposed of. To increase the effectiveness of MSW disposal, the MSW supply chain urgently has to be modified. Ahangar et al. (2021) have proposed a sustainable urban solid waste management system using a mixed integer linear programming model to optimize the number and location of recycling yard construction sites. The system not only takes cost into account, but also includes staffing and optimization of the amount of pollutants. Engelseth (2016) have studied the demonstrates how waste plays a part as a traded product farther down the supply chain in petroleum logistics. From an SCM standpoint, "waste management" is covered along with the "afterlife" of the outbound product and a typical logistical flow. The flow out of the supply base, which is likewise connected to ownership and trading waste products, is not reversed, just the flow from the supply base. Zhou et al. (2021) have developed the TODIM technique for multi-attribute decision making where fundamentally uncertain information is provided in order to use the developed algebra operations and order relations. Ren et al. (2019) have proposed a new distance evaluate of TIFNs and a new ranking technique which consider decision maker attitude behavior, and then develop an extended TODIM decision making methods. Xie et al. (2020) have proposed the number of triangular intuitive fuzzy numbers to the TODIM technique, a mixed intersection of triangular intuitive coefficients and mixed geometric integral operators, and realized a fuzzy information expression and decision-making calculation method suitable for psychology. Kaur et al.

(2022) have suggested the TODIM-based multi criteria decision-making strategy for evaluating inventory policy in a Pythagorean fuzzy environment. In these Pythagorean fuzzy setting, they are considered to reduce this elusiveness by assigning membership and nonmembership roles to these factor. Rangel et al. (2011) have introduced the TODIM effectiveness of this operational research tool was demonstrated through this studies to find a solution to the issue of ranking the services provided in the city in accordance with the standards decided upon by this small business. Autran et al. (2020) have proposed the TOPSIS technique by computing a set of accurate information in pair-wise comparisons related to gain and loss, and proposed these method builds a philosophy-based value function that evaluates the degree to which one option is preferred over another. Make it possible, to demonstrate its effectiveness and validate the results. Zhang (2017) have focused on hesitant qualitative group decision-making issues, and developed a hesitant trapezoidal fuzzy TODIM (interactive and multi-criteria decision making) using a proximity index-based ranking mechanism. Zhang et al (2021) have proposed the TODIM technique based on CPT (CPT–TODIM) for multiple attributes group decision making (MAGDM) with 2TLPFSs. Llamazares (2018) have suggested the TODIM technique and specify the circumstances in which the earlier paradoxes might be avoided. Additionally, the PROMETHEE II method and the simple additive weighting (SAW) method can be produced as special cases for generalized TODIM methods. Leoneti and Gomes (2021) have proposed a new version of the TODIM technique. It fits a multi-criteria method with low implementation complexity and high predictability. It meets several requirements. A technique having little implementation complexity and good predictability. Because it is based on a value function, it is consistent with prospect theory. Passos and Gomes (2014) have introduced TODIM-FSE, a multi-criteria method based on prospect theory for categorizing options. As a result, TODIM-FSE utilizes both the TODIM method and the fuzzy synthetic evaluation strategy. For the purposes of multi-criteria classification, TODIM-FSE employs the novel "contribution" notion. Pei-de (2009) have this paper suggested a new version of the TODIM method. It fits a multi-criteria method with low implementation complexity and high predictability. Because it is based on a value function, it is in line with prospect theory. Lin et al. (2020) have used the conventional TODIM (an interactive acronym in Portuguese). Multi-criteria decision-making technique is extended to handle HFLTS based on the novel Comparison function and distance measurement. Wang et al. (2018) have proposed a 2-tuple linguistic TODIM method that extends the original TODIM method (Portuguese acronym for interactive multi-criteria decision making) for multiple attribute group decision to a linguistic neutrosophic fuzzy environment. Zhan and Huo (2012) have suggested the particle swarm optimization (PSO) technique for scheduling method of cloud computing, through this strategy can shorten the typical work duration and increase resource availability. Da and Xiurun (2005) have used the PSO method with simulated annealing (SA). A better PSO-based artificial neural network (ANN) is created to a better ability to escape from a local optimum and is more efficient than the conservative PSO-based ANN. Liu et al. (2007) have proposed an efficient PSO-based memetic algorithm (MA) for the permutation flow shop scheduling problem (PFSSP) with the goal of minimizing the maximum execution time, which is a polynomial time, typically non-deterministic, (NP) difficult combinatorial optimization problem. Freitas et al. (2020) have identified the Particle Swarm Optimization (PSO) algorithm was inspired by the social and biological behavior of bird flocks searching for food sources. Dings (2017) has suggested a multi-objective particle swarm optimization technique, the simulation outcomes demonstrate that the suggested approach is more efficient in terms of convergence speed and local minimum value, demonstrating that the enhanced PSO technique has a good reference value. Aje and Josephat, (2020) have introduced the concept of PSO, its application to a variety of systems, including power systems, basic PSO changes to improve initial convergence, and combinations with other intelligent systems. It offers algorithms to improve search capability and reduce the time spent getting out of the local optimum. Gad (2022) has presented existing study on method and application published between 2017 and 2019, with technical classification methods for selected content, including actual applications of PSO hybridization, extensions, variants, and algorithms. Shao et al. (2010) have introduced the crossover and mutation operators to improve the particle swarm optimization algorithm called IPSO. On the other hand, a new hybrid algorithmic model is presented, combining the improved PSO technique and the simulated annealing (SA) method. Freitas et al. (2020) have studied the social and biological behavior of bird flocks in search of food sources served as the basis for the (PSO algorithm. Particles in this algorithm inspired by nature move across the search space in pursuit of the global optimum position that minimizes (or maximizes) a particular problem.

5.2 METHODOLOGY

5.2.1 THE ORIGINAL TODIM METHOD

Gomes and Rangel (2009) defined the TODIM method. Assume that F = $\{ f_1, f_2, ..., f_g \}$ represents alternatives, and P = $\{ p_1, p_2, ..., p_h \}$ represents attributes with the weight vector$(v_1,v_2,...,v_h)$ which satisfies $v_t \in [0,1]$ and $\sum_{i=1}^{n} v_t = 1$. Construct a decision matrix.

$$\Psi = \begin{pmatrix} \Psi_{11} & \cdots & \Psi_{1n} \\ \vdots & \ddots & \vdots \\ \Psi_{m1} & \cdots & \Psi_{mn} \end{pmatrix} = (\Psi_{mn})_{g\times h}$$

The calculating steps for MADM based on the TODIM are given as follows:

Step 1. Get a normalized decision matrix $\widetilde{\Psi} = \begin{pmatrix} \breve{\Psi}_{11} & \cdots & \breve{\Psi}_{1n} \\ \vdots & \ddots & \vdots \\ \breve{\Psi}_{m1} & \cdots & \breve{\Psi}_{mn} \end{pmatrix} = (\breve{\Psi}_{mn})_{g\times h}$

Step 2. Calculate the relative weight $\breve{v}_n = \frac{V_n}{\hat{v}}$, n = 1, 2, ……., h,

Where $\hat{v}$ means the maximum amount of the attributes' weights and is used as the reference weight.

Step 3. Calculate the Dominance Degree. The preference index value is computed as follows:

V'n

$$\Phi_n(F_m, F_{m'}) = \begin{cases} \sqrt{\dfrac{\breve{v}n\left(\Phi^*_{mn} - \Phi^*_{m',n}\right)}{\left(\sum_{n=1}^{k} v'_n\right)}}, & \Phi^*_{mn} - \Phi^*_{m',n} > 0, \\ 0, & \Phi^*_{mn} - \Phi^*_{m',n} = 0, \\ \dfrac{-1}{\theta}\sqrt{\dfrac{\left(\sum_{n=1}^{k} v'_n\right)\left|\Phi^*_{mn} - \Phi^*_{m',n}\right|}{v'_n}}, & \Phi^*_{mn} - \Phi^*_{m',n} < 0, \end{cases}$$

where θ is a loss decay coefficient. If θ is smaller, the loss aversion of the decision maker will be higher. $\Phi^*_{mn} - \Phi^*_{m',n}$d emonstrates the gain of alternative Fm to alternative Fm′ if. $\Phi^*_{mn} - \Phi^*_{m',n} > 0$ and the loss if $\Phi^*_{mn} - \Phi^*_{m',n} < 0$. The dominance degree is calculated as follows:

$$\sigma m = \sigma(f_m, f_{m'})\, m = \sum_{n=1}^{h} \Phi n\left(Fm,\ Fm,\right)m,\ \ m' = 1, 2, \ldots, g.$$

Step 4. Compute the overall dominance degree

$$\zeta_m = \frac{\sigma_m - min\sigma_m}{max\sigma_m - min\sigma_m},\ m = 1, 2, \ldots, g.$$

Step 5. Rank the ζ_m (m = 1, 2, …, g), and choose the optimal alternative with the biggestvalue as the optimum decision result.

5.2.2 PARTICLE SWARM OPTIMIZATION (PSO) ALGORITHM

In general, both selection and mutation are contained in PSO. Moreover, no crossover takes place in PSO indicating a higher-mobility in particles with a higher-degree of explorations (Sibalija, 2019). Usually, the motivation of PSO lies in the fish-schooling as well as bird-flocking behavior and represents a population that gets support of non-deterministic optimization-techniques (Sibalija, 2019). Primarily the PSO is utilized for the calculation of the optimum in addition to the nearest optimum-solution by using objective function (fitness function) for the predicament, such as: $f(x) = f(x_1, x_2, x_3, \ldots, x_n)$; where x_i represents the particle's population. By following steps

1. Initialize xi, vi, iterations, pbest, gbest
2. Create random particle (p)
3. For each particles (i)
4. Calculate fitness functions (fi)
5. Update pbest, gbest
6. End for
7. While iterations
8. For each particles I
9. Update vi, xi
10. If Xi > limit then Xi = limit
11. Calculate fitness functions fi
12. Update pbest, gbest
13. End for
14. End while

5.3 RESULTS AND DISCUSSION

5.3.1 OUTPUT OF TODIM METHOD

TABLE 5.1 Criteria and Alternatives for Waste Management in Mining Sector.

Criteria	Alternatives	Key objectives
Air pollution **(Cr-1)**	Release microscopic dust particle **(Al-1)**	Minimization
	Dust from blast operation **(Al-2)**	Minimization
	Smelter operation **(Al-3)**	Minimization
Water pollution **(Cr-2)**	Acid mine drain **(Al-4)**	Minimization
	Lack of portable water **(Al-5)**	Minimization
	Releasing of chemical **(AI-6)**	Minimization
Land pollution **(Cr 3)**	Loss of bio-diversity **(Al-7)**	Minimization
	Sinkholes **(Al-8)**	Minimization
Noise pollution **(Cr-4)**	Crushing **(Al-9)**	Minimization
	Drilling and blasting **(Al-10)**	Minimization
Soil pollution **(Cr -5)**	By heavy metals **(Al-11)**	Minimization
	Other pollutants **(Al-12)**	Minimization

TABLE 5.2 Initial Matrix Given Away.

Weight of criterion	**0.216**	**0.132**	**0.269**	**0.234**	**0.149**
Kind of criterion	**–-1**	**–1**	**–1**	**–1**	**1**
	C-1	**C-2**	**C-3**	**C-4**	**C-5**
A-1	65	39	17.9	55	65
A-2	36	39	18.9	66	42
A-3	23	39	21.1	22	12
A-4	59	51	35.2	96	26
A-5	66	49	18.4	55	95
A-6	25	50	21.7	66	84
A-7	95	41	24.9	24	26
A-8	45	41	25.9	66	45
A-9	55	41	29.2	96	29
A-10	55	53	39.4	23	25
A-11	22	53	20.4	87	26
A-12	33	53	23.1	26	23
Sum of square	579	549	296.1	682	498

TABLE 5.3 Normalized Matrix Given Away.

Weight of criterion	**0.216**	**0.132**	**0.269**	**0.234**	**0.149**
Kind of criterion	**−1**	**−1**	**−1**	**−1**	**1**
	C-1	**C-2**	**C-3**	**C-4**	**C-5**
A-1	0.0507	0.0961	0.1083	0.0642	0.1305
A-2	0.0915	0.0961	0.1026	0.0535	0.0843
A-3	0.1432	0.0961	0.0919	0.1605	0.0241
A-4	0.0558	0.0735	0.0551	0.0368	0.0522
A-5	0.0499	0.0765	0.1054	0.0642	0.1908
A-6	0.1317	0.0750	0.0894	0.0535	0.1687
A-7	0.0347	0.0914	0.0779	0.1471	0.0522
A-8	0.0732	0.0914	0.0749	0.0535	0.0904
A-9	0.0599	0.0914	0.0664	0.0368	0.0582
A-10	0.0599	0.0707	0.0492	0.1535	0.0502
A-11	0.1497	0.0707	0.0951	0.0406	0.0522
A-12	0.0998	0.0707	0.0839	0.1358	0.0462
Recalculated weights	0.8030	0.4907	1	0.8699	0.5539

5.3.2 OUTPUT OF PARTICLE SWARM OPTIMIZATION (PSO)

Particle swarm optimization procedure is made on information that particles travel through search spaces with the speeds that are enthusiastically acclimated by their classical conduct. Thus, the particles can move toward the improved and better search-zone over the way of the pursuit procedure. PSO methodology has a huge numbers of postulates. At present, there is numerous kind of created improvement strategy of molecule swarm, which has generally excellent capability of tackling huge scale enhancement issues. In the present analysis, 28 of iterations with best function value to be −0.0571 were obtained in order to minimize the objective function for minimizing the pollution that is caused by the mining sector (Figure 5.1).

The regression equation is

Pollution = −0.192 – 0.266 Release microscopic dust particle
−0.0312 Dust from blast operation + 0.0856 Smelter operation
+ 0.0536 Acid mine drain – 0.0014 Lack of portable water
−0.0482 releasing of chemical + 0.0497 Loss of bio-diversity
+0.0511 Sinkholes + 0.162 Crushing – 0.0751 Drilling and blasting
+0.0714 By heavy metals + 0.175 Other pollutants

TABLE 5.4 Sum of Single Criterion Dominances [δ (ai,ak)].

	A1	A2	A3	A4	A5	A6	A7	A8	A9	A10	A11	A12
A-1	0	–0.2625	–1.1036	0.2080	–0.5439	–0.9444	–0.3131	–0.0758	0.1084	–0.5310	–0.3771	–0.7792
A-2	–0.8227	0	–1.0171	0.3872	–1.0151	–1.0715	–0.3461	–0.0270	0.3311	–0.3221	–0.2921	–0.5852
A-3	–0.8008	–0.5713	0	0.0273	–0.9386	–0.6981	–0.1390	–0.2931	–0.0667	–0.0790	–0.4908	–0.1080
A-4	–1.8928	–1.9722	–2.0823	0	–1.8540	–2.2067	–1.2787	–1.6966	–0.9119	–0.7672	–1.1531	–1.3802
A-5	–0.4550	–0.6210	–1.4665	0.1944	0	–0.4283	–0.6445	–0.4017	–0.2283	–0.5375	–0.3814	–0.7835
A-6	–0.6719	–0.4168	–1.2572	0.4323	–0.8177	0	–0.6536	–0.0702	0.0409	–0.2688	–0.2235	–0.3131
A-7	–1.3828	–1.3209	–1.3000	–0.0253	–1.3660	–1.5665	0	–0.7519	–0.3265	–0.3496	–0.7723	–0.5656
A-8	–1.2044	–0.7706	–1.5863	0.3208	–1.2562	–1.4311	–0.5716	0	0.2331	–0.3875	–0.6865	–0.9944
A-9	–1.5775	–1.6237	–1.7732	0.1634	–1.5752	–1.9505	–0.7894	–1.1572	0	–0.5514	–1.0165	–1.2411
A-10	–1.4524	–1.5923	–1.5684	–0.2139	–1.4464	–1.8811	–0.7261	–1.3193	–0.7158	0	–1.0112	–0.7004
A-11	–1.5571	–1.1932	–1.0231	0.1310	–1.5403	–1.1969	–0.8452	–0.9347	–0.3402	–0.4271	0	–0.4495
A-12	–1.2596	–1.0268	–1.3263	–0.0082	–1.2433	–1.4736	–0.6580	–0.6765	–0.3666	–0.2498	–0.7357	0

TABLE 5.5 Final Ranking.

Alternatives	Global dominance G(ai)	Relative overall value V(ai)	Ranking
A-1	–4.6141	0.9650	3
A-2	–4.7806	0.9522	4
A-3	–4.1580	1	1
A-4	–17.1958	0	12
A-5	–5.7533	0.8776	5
A-6	–4.2196	0.9953	2
A-7	–9.7275	0.5728	9
A-8	–8.3348	0.6796	6
A-9	–13.0924	0.3147	11
A-10	–12.6271	0.3504	10
A-11	–9.3764	0.5998	8
A-12	–9.0245	0.6267	7

Min = −17.1958
Max = −4.1580

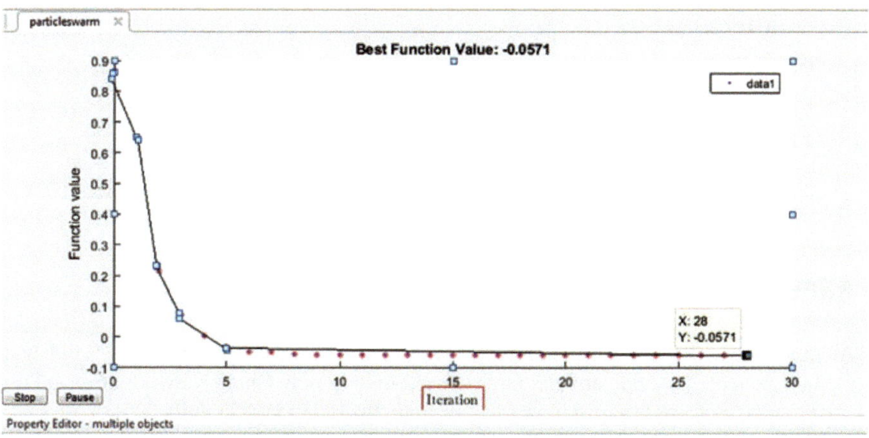

FIGURE 5.1 Result of particle swarm optimization (PSO).

5.4 CONCLUSION

The element influencing waste management in the mining sector in order to prevent pollution. Therefore, during this study a significant step has been taken in synthesizing some of the important risk factors and their ranking, such that possible preventive measures can be initiated. The results from overall ranking of all the five criteria and their 12 alternatives considered were found as: Smelter operation 1st, Releasing of chemical 2nd, Release microscopic dust particle 3rd, Dust from blast operation 4th, Lack of portable water 5th, Sinkholes 6th, Other pollutants 7th, By heavy metals 8th, Loss of bio-diversity 9th, Drilling and blasting 10th, Crushing 11th, and Acid mine drain 12th. Particle swarm optimizations (PSO) is also used (Figure 5.1). "Optimization completed: relative change in the objective value..." particle swarm reached the value 0.057100 after 3292 function evaluations, with 28 iterations yielding the best function value of 0.0571 in order to minimize the objective function for minimizing the pollution caused by the mining sector.

KEYWORDS

- **management**
- **pollution**
- **affecting**
- **technique**
- **effectiveness**
- **minimization**

REFERENCES

Ahangar, S. S.; Sadati, A.; Rabbani, M. Sustainable Design of a Municipal Solid Waste Management System in an Integrated Closed-loop Supply Chain Network Using a Fuzzy Approach: A Case Study. *J. Indus. Prod. Eng.* **2021,** *38* (5), 323–340.

Aje, O. F.; Josephat, A. A. The Particle Swarm Optimization (PSO) Algorithm Application–A Review. *Glob. J. Eng. Technol. Adv.* **2020,** *3* (3), 001–006.

Amemba, C. S.; Nyaboke, P. G.; Osoro, A.; Mburu, N. Elements of Green Supply Chain Management. *Eur. J. Bus. Manag.* **2013,** *5* (12), 51–61.

Autran, A. S. S. L. F.; Gomes, M.; Vijayalakshmi, K. R. Assessment of MCDM Problems by TODIM Using Aggregated Weights. *Neutrosophic Sets Syst.* **2020,** *78*.

Björklund, T.; Fors, W. Waste Management With a Green Supply Chain: A Case Study Regarding How for-Profit Organisations Should Utilise Waste Management. 2018.

Da, Y.; Xiurun, G. An Improved PSO-based ANN with Simulated Annealing Technique. *Neurocomputing* **2005,** *63,* 527–533.

Ding, Z. Research of Improved Particle Swarm Optimization Algorithm. In *AIP Conference Proceedings*, Vol. 1839, No. 1; AIP Publishing LLC, 2017; p 020148.

Engelseth, P. Developing the Service Value: A Case Study of Waste Management in Offshore Petroleum Logistics. *WIT Trans. Ecol. Environ.* **2016,** *202,* 135–146.

Freitas, D.; Lopes, L. G.; Morgado-Dias, F. Particle Swarm Optimisation: A Historical Review up to the Current Developments. *Entropy* **2020,** *22* (3), 362.

Gad, A. G. Particle Swarm Optimization Algorithm and Its Applications: A Systematic Review. *Arch. Comput. Meth. Eng.* **2022,** 1–31.

Gomes, L.; Rangel, L. A. D. An Application of the TODIM Method to the Multi-criteria Rental Evaluation of Residential Properties. *Eur. J. Oper. Res.* **2009,** *193* (1), 204–211.

Hicks, C.; Heidrich, O.; McGovern, T.; Donnelly, T. A Functional Model of Supply Chains and Waste. *Int. J. Prod. Econ.* **2004,** *89* (2), 165–174.

Hui, R. K. Green Supply Chain Management: A Survey Based Paper. *Int. J. Eng. Res. Technol.* **2021,** *10* (4).

Kaur, P.; Pradhan, B. L.; Priya, A. TODIM Approach for Selection of Inventory Policy in Supply Chain. *Math. Probl. Eng.* **2022.**

Leoneti, A. B.; Gomes, L. F. A. M. A Novel Version of the TODIM Method based on the Exponential Model of Prospect Theory: The ExpTODIM Method. *Eur. J. Oper. Res.* **2021,** *295* (3), 1042–1055.

Lin, M.; Wang, H.; Xu, Z. TODIM-based Multi-criteria Decision-making Method with Hesitant Fuzzy Linguistic Term Sets. *Artif. Intell. Rev.* **2020,** *53* (5), 3647–3671.

Liu, B.; Wang, L.; Jin, Y. H. An Effective PSO-Based Memetic Algorithm for Flow Shop Scheduling. *IEEE Transac. Syst. Man Cybernet. Part B (Cybernetics)* **2007,** *37* (1), 18–27.

Llamazares, B. An Analysis of the Generalized TODIM Method. *Eur. J. Oper. Res.* **2018,** *269* (3), 1041–1049.

Mandičák, T.; Mésároš, P.; Spišáková, M. Impact of Information and Communication Technology on Sustainable Supply Chain and Cost Reducing of Waste Management in Slovak Construction. *Sustainability* **2021,** *13* (14), 7966.

Nyarko, J.; Yeboah, F. A.; Yeboah, F. A.; Larbi, I. S.; Osei-Bonsu, M. Green Supply Chain in Solid Waste Management: Case Study of EcoCare H2H Waste Collection, Goaso, Ghana. In *Green Supply Chain-Competitiveness and Sustainability*; IntechOpen, 2020.

Passos, C. A.; Autran Monteiro Gomes, L. F. TODIM-FSE: A Multi-criteria Classification Method based on Prospect Theory. *Mult. Criteria Decis. Mak.* **2014,** *9,* 123–139.

Pei-de, L. A Novel Method for Hybrid Multiple Attribute Decision Making. *Knowl.-Based Syst.* **2009,** *22* (5), 388–391.

Rangel, L. A. D.; Gomes, L. F. A. M.; Cardoso, F. P. An Application of the TODIM Method to the Evaluation of Broadband Internet Plans. *Pesqui. Oper.* **2011,** *31,* 235–249.

Ren, H.; Liu, M.; Zhou, H. Extended TODIM Method for MADM Problem Under Trapezoidal Intuitionistic Fuzzy Environment. *Int. J. Comp. Commun. Control* **2019,** *14* (2), 220–232.

Saif, Y.; Rizwan, M.; Almansoori, A.; Elkamel, A. A Circular Economy Solid Waste Supply Chain Management Based Approach Under Uncertainty. *Energy Procedia* **2017,** *142,* 2971–2976.

Shao, Y.; Chen, Q.; Li, C. Research on Hybrid Improved PSO Algorithm. In *International Symposium on Intelligence Computation and Applications*; Springer: Berlin, Heidelberg, 2010; pp 234–242.

Sibalija, P. Particle Swarm Optimisation in Designing Parameters of Manufacturing Processes: A Review (2008–2018). *Appl. Soft Comput.* **2019,** *84,* 1057.

Wang, J.; Wei, G.; Lu, M. TODIM Method for Multiple Attribute Group Decision Making Under 2-Tuple Linguistic Neutrosophic Environment. *Symmetry* **2018,** *10* (10), 486.

Xie, Y.; Li, G.; Zhou, X. In *Research on the Improvement of TODIM Decision Method Based on Triangular Intuitionistic Fuzzy Information*, Proceedings of the 4th International Conference on Computer Science and Application Engineering, 2020; pp 1–6.

Xin, C.; Wang, J.; Wang, Z.; Wu, C. H.; Nawaz, M.; Tsai, S. B. Reverse Logistics Research of Municipal Hazardous Waste: A Literature Review. *Environ. Dev. Sustain.* **2021,** 1–37.

Xu, X.; Yang, Y. Municipal Hazardous Waste Management with Reverse Logistics Exploration. *Energy Rep.* **2022,** *8,* 4649–4660.

Yang, X.; Guo, X.; Yang, K. Redesigning the Municipal Solid Waste Supply Chain Considering the Classified Collection and Disposal: A Case Study of Incinerable Waste in Beijing. *Sustainability* **2021,** *13* (17), 9855.

Zhan, S.; Huo, H. Improved PSO-based Task Scheduling Algorithm in Cloud Computing. *J. Informa. Comput. Sci.* **2012,** *9* (13), 3821–3829.

Zhang, X. A Closeness Index-based TODIM Method for Hesitant Qualitative Group Decision Making. *Informatica* **2017,** *28* (3), 565–581.

Zhang, Y.; Wei, G.; Guo, Y.; Wei, C. TODIM Method Based on Cumulative Prospect Theory for Multiple Attribute Group Decision-making Under 2-tuple Linguistic Pythagorean Fuzzy Environment. *Int. J. Intell. Syst.* **2021,** *36* (6), 2548–2571.

Zhou, Z.; Xuan, K.; Tao, Z.; Zhou, L. Generalized-TODIM Method for Multi-criteria Decision Making with Basic Uncertain Information and its Application. , 2021, arXiv preprint arXiv:2104.11597.

CHAPTER 6

Supplier Selection Criterion in the Mining Industry for Sustainable Supply Chain Management

ABSTRACT

Supplier threat evaluation used to be developed in response to these realizations, and is leveraged to identify, understand, and mitigate doable risks that can influence a business, motive losses, or impact partnership. Inefficient spend administration or processes, inaccurate data, low-quality merchandise and services, and lack of compliance can lead to principal challenges for corporations in the long-run, along with a decline in quality, cost-efficiency, sustainability, and profitability. To preserve a streamlined, resilient, and obvious provide chain, groups need to be equally conscious of their present day or plausible partners' organizational structure, processes, and operations, and recognize any manageable dangers as properly as their impact. Hence, supplier selection criteria are prioritized.

6.1 INTRODUCTION

Indian mines are the reason for the industrial development. Many other developing countries are also dependent on mines for their industrialization. India is rich in fuels and metals. Still it is behind compared to other countries. Due to lack of advance and sustainable technology for mining extraction, transportation, and utilization, many mines are unutilized and are non-functioning. Although Indian Government is trying a lot to

Sustainable Supply Chain Management in the Mining Industry.
Suchismita Satapathy and Hullash Chauhan (Authors)

develop and support mining industries, still there is a lot to be done. The challenges are creating obstacles for the rapid growth of mines.

As per statistics, Indian mines are filled with 85 types of minerals and 3500 mines (i.e., coal, lignite, iron ore, bauxite, manganese, chromite, copper, zinc and lead, limestone, phosphorite, diamond, etc.) around 317 K hectare mining's areas within 23 states. India is the second-largest producer and importer of coal in the world. Mining industry has many challenges like land acquisitions, habitation issues, and the most important is its effect on the environment. To enhance the growth of coal mines, Govt. has allowed private coal sectors for mining. Companies like Leighton and Rio Tinto, etc., had tried and invested in coal sector but struggled a lot and unable to face the challenges and lastly lost interest in India mining.

Lack of political stability, changes of policies, laws and regulations in environmental condition, lack of conjunction between state government and central government have also contributed to some extent and it is one of the reasons for slow growth of the industry.

After 2014, the Government has been trying to reform the mining sectors by giving auctions of mining lease, creating District Mineral Foundation (DMF), forming National Mineral Exploration Trust (NMET) and by providing rating to the mining leases on sustainable development framework (SDF), adding detection system for mining surveillance to avoid illegal mining activities, are the positive moves in mining sector. Few revisions were done for national mineral policy to establish sustainability development goals. Illegal land acquisition and mass displacement of natives to other locations are the most important challenge of Indian mining industries. Forest and environment challenge is the second most important challenge mining sector is facing. Apart of them, high capital investment, latest technology and modernization, and digitalization in mining are also effecting mining sector.

To support the rapid growth of mining industry, the Government and Industry are taking a lot of steps. But selection of supplier is also very important factor to produce clean minerals. So some factors must be considered like quality timeliness, pollution free delivery, etc.

Value for money, quality, reliability, and service are mostly important points to select suppliers. Rezaei et al. (2020) have explored that supplier selection is a critical process in every organization, offering quantity discount is an important feature in the selection of the best suppliers. Alegoz and Yapicioglu (2019) have discussed about quantity discount and provided low cost while allocating large volume orders to the suppliers.

Arabsheybani et al. (2018) have found that factors such as quality, cost, and delivery time should be considered in the supplier selection problem. As per the various studies (Cheraghalipour and Farsad, 2018; Moheb-Alizadeh and Handfield, 2019), supplier selection is a multi-criteria decision-making (MCDM) problem. Çebi and Otay (2016) and Arabsheybani et al. (2018) have explored that cost saving and minimization of risks can be achieved by using suitable supplier selection methods. Hamdan and Cheaitou (2017a) have found that supplier selection and order allocation are very important in green supply chain management considering sustainability and environmental factors. Hamdan and Cheaitou (2017b) and Ghadimi et al. (2018) have found sustainable supplier selection encompasses cost, environmental and social criteria and supplier's performance history.

TABLE 6.1 Supplier Selection Criteria.

Sl. no	Selection criteria
1.	Quality
2.	Delivery
3.	Price
4.	Performance history
5.	Technical capacity
6.	Impression
7.	Geographical location
8.	Operational controls
9.	Production facilities and capacity
10.	Reputation and position in industry
11.	Financial position
12.	Management and organization
13.	Warranties and claims
14.	Occupational health management
15.	Environmental management
16.	Training aids
17.	Amount of past business
18.	Reciprocal arrangements
19.	Relationship closeness
20.	Labor relations
21.	Packaging ability
22.	Attitude
23.	After sales services

TABLE 6.1 *(Continued)*

Sl. no	Selection criteria
24.	Conflict resolution
25.	Occupational safety management
26.	Social acceptability

Source: Dickson (1966)

TABLE 6.2 Sustainable Supplier Selection Criteria.

	Environmental criteria	Social criteria	Economic criteria
1.	Environmental management system	Stakeholder engagement	Quality
2.	Resource consumption	Staff training	Flexibility
3.	Ecological design	Commitment in social management	Price
4.	Recycling	Commitment to health and safety management	Delivery term
5.	Ecological impact control	Stake holder relation	Relationship
6.	Sewage water	Code of social conduct	Cost
7.	Energy consumption	Donations for sustainable products	Technical capacity
8.	Reuse	Rights of interested parties	Logistical cost
9.	Air emissions	Annual number of accidents	Reverse logistic
10.	Environmental code of conduct	Safety practices	Rejection rate
11.	Green Image	Rights of stakeholders	Professionalism
12.	Pollution control	Human rights	Product development
13.	Greenhouse emission	Health and safety of employees	Process improvement
14.	Green transportation	Respect for the policy	JIT
15.	Hazardous material management	Information disclosure	Commitment
16.	Noise	Enterprise Resource Planning	Integrity

Source: Zimmer et al. (2016)

6.2 RESEARCH METHODOLOGY

A survey is conducted on Indian mining Industries like coal, limestone, quartz, iron, copper, bauxite and gold, etc., by designing a standard questionnaire selected from literature review and considering opinion of experts from mining industry. The questionnaire is designed by considering both supplier selection criteria and sustainable supplier selection criteria. A total of 120 questions are circulated by personal contact and internet, among them responses received are 105 in number. Data are collected by using Likert scale (1–5) (i.e., 1. disagree, 2. somewhat disagree, 3. no opinion, 4. somewhat disagree, and 5. disagree).Then data are analyzed by considering the parameters of supplier selection criterion by WASPAS method.

6.3 RESULT AND DISCUSSION

WASPAS method is used for finding or selecting best supplier selection criterion followed in all mining sectors for supplier selection in both Sustainable supplier selection and supplier selection criteria.

Steps of WASPAS method are as follows below.

Step 1: Decision matrix is constituted.

The decision matrix (X), which illustrates how options perform in relation to criteria, is constructed in the first stage. This matrix matches that of eq 6.1 given earlier. X_{ij} in the choice matrix denotes how well an alternative performed in relation to j criterion.

Step 2: Decision matrix is normalized.

All of the components of the decision matrix are normalized using the following two equations so that the performance matrices are similar and dimensionless:

$$\bar{X} = \frac{X_{ij}}{Max\ i\ X_{ij}} \text{ for beneficial criteria} \tag{6.1}$$

$$\bar{X}_j = \frac{Min\ i\ X_{ij}}{X_{ij}} \text{ for non-beneficial criteria} \tag{6.2}$$

Where X_{ij} normalized value is represented by $\bar{X}_{ij}$.

Step 3: Relative performances of alternatives are calculated according to the WSM method.

In the WASPAS technique, a joint optimality criterion is sought based on two optimality criteria. The first optimality criterion, which is a mean weighted success criterion, is comparable to the WSM technique. It is a well-liked and approved MCDM method used to assess many alternatives in relation to a set of decision criteria. The overall relative importance of the ith alternative is determined using the WSM technique (MacCrimon, 1968; Triantaphyllou and Mann, 1989) as follows:

$$\mathbf{O_i}^{(1)} = \sum\nolimits_{j=1}^{n} \bar{X}_{ij} W_j \tag{6.3}$$

Where, *Wj* is the weight (relative weight) of the jth criterion.

Step 4: Relative performances of alternatives are calculated according to the WPM method.

On either hand, the WPM technique uses the following eq 6.4 to determine the overall relative value of each alternative: To get this, multiply the alternative value for each criterion by the power of the same indexed criterion weight.

$$Q_i^{(1)} = \prod\nolimits_{j=1}^{n} (\bar{X}_{ij}) W_{ij} \tag{6.4}$$

The following is then suggested as a combined generalized criterion of weighted aggregation of additive and multiplicative approaches.

Step 5: Final performance of alternatives is calculated.

The relative performances derived by Eqs 6.3 and 6.4 are added to determine the final performance of the alternatives Pi which illustrates the positions of the alternatives in a general ranking.

$$\mathbf{Q_i = 0.5\ P_i^{(1)} + 0.5\ P_i^{(2)}} \tag{6.5}$$

The above equation can be used to improve decision-making accuracy and ranking efficiency. That is a parameter that accepts values in the range of 0–1.

$$\mathbf{Q_i = \alpha\ P_i\ (1) + (1 - \alpha)\ P_i^{(2)}} \tag{6.6}$$

When is equal to 1, the WASPAS method becomes the WSM method, and when is equal to 0, it becomes the WPM method. The user has the complete control over the value.

Step 6: Alternatives are ranked.

The final ranks of the alternatives are determined by sorting Pi values from highest to lowest. The first alternative in this list is the best, while the last alternative is the worst.

TABLE 6.3 Decision Matrix of Supplier Selection Criteria in Mining.

	Quartzite	Lime stone	Coal	Iron	Copper	Gold	Bauxite
Quality	3.555555556	3.916666667	3.583333333	3.666666667	3.923076923	3.846153846	3.545454545
Delivery	3.888888889	3.833333333	3.833333333	3.583333333	3.846153846	3.923076923	3.818181818
Price	4.111111111	3.5	3.916666667	3.5	3.461538462	4.076923077	4
Performa nce history	3.888888889	2.833333333	3.5	3	2.923076923	3.461538462	3.727272727
Technical capacity	4	3.833333333	3.5	4.166666667	2.923076923	2.692307692	3.636363636
Impression	4.666666667	3.666666667	4	3.75	3.461538462	3.307692308	4.090909091
Geographical location	3.555555556	3.416666667	3.5	3.25	3	3.615384615	3.454545455

TABLE 6.3 *(Continued)*

	Quartzite	Lime stone	Coal	Iron	Copper	Gold	Bauxite
Operational controls	3.666666667	2.916666667	3.416666667	3.083333333	3.692307692	3.307692308	3.272727273
Production facilities and capacity	3.444444444	4	3.166666667	4.083333333	3.769230769	3	3.272727273
Reputation and position in industry	3.666666667	3	3.666666667	3	3.769230769	3.230769231	3.727272727
Financial position	4.222222222	3.25	3.75	3.583333333	3.230769231	3.076923077	3.909090909
Management and organization	4	3.083333333	3.666666667	3.083333333	3.692307692	3.076923077	3.909090909
Warranties and claims	4.444444444	3.25	4.5	3.5	3.615384615	2.923076923	4.454545455
Occupational health management	4	3.583333333	4	3.666666667	2.461538462	3	4
Environ mental management	4.222222222	3.833333333	4.166666667	3.75	2.923076923	3.615384615	4.272727273

TABLE 6.3 *(Continued)*

	Quartzite	Lime stone	Coal	Iron	Copper	Gold	Bauxite
Training aids	3.75	3.583333333	3	3.21	3.34	3.384615385	3.727272727
Amount of past business	4.111111111	3.416666667	4.166666667	3.25	3.153846154	3	4.272727273
Reciprocal arrangements	3.888888889	3.25	4.166666667	3.333333333	3.076923077	2.923076923	4.272727273
	3.888888889	3.666666667	3.833333333	3.75	3.076923077	3.384615385	3.818181818
	3.888888889	3.916666667	4	4.166666667	3.615384615	3.692307692	4
Relationship closeness	3.888888889	3.833333333	4	3.75	3.923076923	3.846153846	4
Labor relations	4	3.25	4.083333333	3.416666667	3.461538462	3.692307692	4

TABLE 6.3 *(Continued)*

	Quartzite	Lime stone	Coal	Iron	Copper	Gold	Bauxite
	4.222222222	4	4.166666667	4	3.230769231	3.538461538	4.272727273
Packaging ability	3.777777778	3.416666667	3.75	3.416666667	3.923076923	2.307692308	3.727272727
Attitude	4.111111111	3.916666667	4.25	3.75	4	2.769230769	4.181818182
After sales services	4.222222222	3.416666667	4.083333333	3.583333333	3.615384615	3.153846154	4.090909091

TABLE 6.4 Decision Matrix is Normalized.

	Quartzite	Lime stone	Coal	Iron	Copper	Gold	Bauxite
Quality	0.761904762	0.979166667	0.796296296	0.818181818	0.627450981	0.6	0.705128205
Delivery	0.833333333	0.958333333	0.851851852	0.837209302	0.64	0.588235294	0.654761905
Price	0.880952381	0.875	0.87037037	0.857142857	0.711111111	0.566037736	0.625
Performance history	0.833333333	0.708333333	0.777777778	1	0.842105263	0.666666667	0.670731707
Technical capacity	0.857142857	0.958333333	0.777777778	0.72	0.842105263	0.857142857	0.6875
Impression	1	0.916666667	0.888888889	0.8	0.711111111	0.697674419	0.611111111
Geographical location	0.761904762	0.854166667	0.777777778	0.923076923	0.820512821	0.638297872	0.723684211
Operational	0.78571	0.7291	0.759259	0.9729	0.666	0.697674419	0.763888889

TABLE 6.4 *(Continued)*

	Quartzite	Lime stone	Coal	Iron	Copper	Gold	Bauxite
controls	4286	66667	259	72973	666667		
Production facility and capacity	0.738095238	1	0.703703704	0.734693878	0.653061225	0.769230769	0.763888889
Reputation and position in industry	0.785714286	0.75	0.814814815	1	0.653061225	0.714285714	0.670731707
Financial position	0.904761905	0.8125	0.833333333	0.837209302	0.761904762	0.75	0.639534884
Management and organization	0.857142857	0.770833333	0.814814815	0.972972973	0.666666667	0.75	0.639534884
Warranties and claims	0.952380952	0.8125	1	0.857142857	0.680851064	0.789473684	0.56122449
Occupational health management	0.857142857	0.895833333	0.888888889	0.818181818	1	0.769230769	0.625

TABLE 6.4 *(Continued)*

	Quartzite	Lime stone	Coal	Iron	Copper	Gold	Bauxite
Environmental management	0.904761905	0.958333333	0.925925926	0.8	0.842105263	0.638297872	0.585106383
Training aids	0.803571429	0.895833333	0.666666667	0.934579439	0.736987563	0.681818182	0.670731707
Amount of past business	0.880952381	0.854166667	0.925925926	0.923076923	0.780487805	0.769230769	0.585106383
Reciprocal arrangements	0.833333333	0.8125	0.925925926	0.9	0.8	0.789473684	0.585106383
Relationship closeness	0.833333333	0.958333333	0.888888889	0.8	0.627450981	0.6	0.625
Labor relations	0.857142857	0.8125	0.907407407	0.87804878	0.711111111	0.625	0.625
Packaging ability	0.809523809	0.854166667	0.833333333	0.87804878	0.627450981	1	0.670731707

TABLE 6.4 *(Continued)*

	Quartzite	Lime stone	Coal	Iron	Copper	Gold	Bauxite
Attitude	0.880952381	0.979166667	0.944444444	0.8	0.615384616	0.833333333	0.597826087
After sales services	0.904761905	0.854166667	0.907407407	0.837209302	0.680851064	0.731707317	0.611111111
Conflict resolution	0.616071429	0.953125	0.972222222	0.631578947	0.587830084	0.527472528	0.555555556
Occupational safety management	0.897321429	0.796875	0.652777778	0.857142857	0.772247361	0.971659919	1
Social acceptability	0.869048571	0.972225	0.9506	0.76057195	0.590764505	0.594306543	0.625
Reciprocal arrangements	0.833333333	0.8125	0.925925926	0.9	0.8	0.789473684	0.585106383
Relationship closeness	0.833333333	0.958333333	0.888888889	0.8	0.627450981	0.6	0.625

TABLE 6.4 *(Continued)*

	Quartzite	Lime stone	Coal	Iron	Copper	Gold	Bauxite
Labor relations	0.857142857	0.8125	0.907407407	0.87804878	0.711111111	0.625	0.625
Packaging ability	0.809523809	0.854166667	0.833333333	0.87804878	0.627450981	1	0.670731707
Attitude	0.880952381	0.979166667	0.944444444	0.8	0.615384616	0.833333333	0.597826087
After sales services	0.904761905	0.854166667	0.907407407	0.837209302	0.680851064	0.731707317	0.611111111
Conflict resolution	0.616071429	0.953125	0.972222222	0.631578947	0.58783008	0.527472528	0.555555556
Occupational safety management	0.897321429	0.796875	0.652777778	0.857142857	0.772247361	0.971659919	1
Social acceptability	0.869048571	0.972225	0.9506	0.76057195	0.590764505	0.594306543	0.625

TABLE 6.5 Weighted Matrixes for Supplier Selection.

	Quartzite	Lime stone	Coal	Iron	Copper	Gold	Bauxite
	0.25	0.25	0.25	0.25	0.25	0.25	0.25
Quality	0.190476	0.244791	0.199074	0.204545	0.156862		0.176282
	19	667	074	455	745	0.15	051
Delivery	0.208333	0.239583	0.212962	0.209302		0.147058	0.163690
	333	333	963	326	0.16	824	476
Price	0.220238		0.217592	0.214285	0.177777	0.141509	
	095	0.21875	593	714	778	434	0.15625
Performance history	0.208333	0.177083	0.194444		0.210526	0.166666	0.167682
	333	333	444	0.25	316	667	927
Technical	0.214285	0.239583	0.194444		0.210526	0.214285	
capacity	714	333	444	0.18	316	714	0.171875
Impression		0.229166	0.222222		0.177777	0.174418	0.152777
	0.25	667	222	0.2	778	605	778
Geographical	0.190476	0.213541	0.194444	0.230769	0.205128	0.159574	0.180921
location	19	667	444	231	205	468	053
Operational controls	0.196428	0.182291	0.189814	0.243243	0.166666	0.174418	0.190972
	571	667	815	243	667	605	222
Production facilities	0.184523		0.175925	0.183673	0.163265	0.192307	0.190972
and capacity	81	0.25	926	469	306	692	222
Reputation and	0.196428		0.203703		0.163265	0.178571	0.167682
position in industry	571	0.1875	704	0.25	306	429	927

TABLE 6.5 *(Continued)*

	Quartzite	Lime stone	Coal	Iron	Copper	Gold	Bauxite
Financial position	0.226190	0.203125	0.208333	0.209302	0.190476	0.1875	0.159883
	476		333	326	191		721
Management and organization	0.214285	0.192708	0.203703	0.243243	0.166666	0.1875	0.159883
	714	333	704	243	667		721
Warranties and claims	0.238095	0.203125	0.25	0.214285	0.170212	0.197368	0.140306
	238			714	766	421	122
Occupational health management	0.214285	0.223958	0.222222	0.204545	0.25	0.192307	0.15625
	714	333	222	455		692	
Environmental management	0.226190	0.239583	0.231481	0.2	0.210526	0.159574	0.146276
	476	333	481		316	468	596
Training aids	0.200892	0.223958	0.166666	0.233644	0.184246	0.170454	0.167682
	857	333	667	86	891	545	927
Amount of past business	0.220238	0.213541	0.231481	0.230769	0.195121	0.192307	0.146276
	095	667	481	231	951	692	596
Reciprocal arrangements	0.208333		0.231481			0.197368	0.146276
	333	0.203125	481	0.225	0.2	421	596
	0.208333	0.229166	0.212962			0.170454	0.163690
	333	667	963	0.2	0.2	545	476
	0.208333	0.244791	0.222222		0.170212		
	333	667	222	0.18	766	0.15625	0.15625

TABLE 6.5 *(Continued)*

	Quartzite	Lime stone	Coal	Iron	Copper	Gold	Bauxite
Relationship closeness	0.208333	0.239583	0.222222		0.156862		
	333	333	222	0.2	745	0.15	0.15625
Labor	0.214285	0.203125	0.226851	0.219512	0.177777	0.15625	0.15625
Relations	714		852	195	778		
	0.226190		0.231481		0.190476	0.163043	0.146276
	476	0.25	481	0.1875	191	478	596
Packaging ability	0.202380	0.213541	0.208333	0.219512	0.156862		0.167682
	952	667	333	195	745	0.25	927
Attitude	0.220238	0.244791	0.236111		0.153846	0.208333	0.149456
	095	667	111	0.2	154	333	522
After sales services	0.226190	0.213541	0.226851	0.209302	0.170212	0.182926	0.152777
	476	667	852	326	766	829	778
Conflict resolution	0.154017	0.238281	0.243055	0.157894	0.146957	0.131868	0.138888
	857	25	556	737	52	132	889
Occupational safety	0.224330	0.199218	0.163194	0.214285	0.193061	0.242914	
management	357	75	444	714	84	98	0.25
Social acceptability	0.217262	0.243056		0.190142	0.147691	0.148576	
	143	25	0.23765	988	126	636	0.15625

TABLE 6.6 Ranking of Supplier Selection Criteria.

	Q1	Q2	Q3	Rank
Quality	1.322032182	0.052974304	0.687503243	
Delivery	1.340931255	0.054104327	0.697517791	
Price	1.346403614	0.054511757	0.700457685	
Performance history	1.37473702	0.056993981	0.715865501	
Technical capacity	1.425000522	0.061095264	0.743047893	1st
Impression	1.406363049	0.058933177	0.732648113	2nd
Geographical location	1.374855258	0.057338148	0.716096703	
Operational controls	1.34383579	0.055041281	0.699438535	
Production facilitiesand capacity	1.340668426	0.054677338	0.697672882	
Reputation and position in industry	1.347151937	0.055028148	0.701090042	
Financial position	1.384811047	0.058157293	0.72148417	3rd
Management and organization	1.367991382	0.056545353	0.712268368	
Warranties and claims	1.413393262	0.059115101	0.736254181	
Occupational health management	1.463569417	0.063624387	0.763596902	
Environmental management	1.413632671	0.059255744	0.736444207	
Training aids	1.34754708	0.055113574	0.701330327	
Amount of past business	1.429736713	0.060937792	0.745337253	
Reciprocal arrangements	1.411584832	0.059727709	0.73565627	
Relationship closeness	1.333251634	0.053439943	0.693345788	
Labor relations	1.354052539	0.055382018	0.704717279	
	1.394968222	0.057784679	0.726376451	

TABLE 6.6 *(Continued)*

	Q1	Q2	Q3	Rank
Packaging ability	1.418313819	0.060041922	0.739177871	
Attitude	1.412776882	0.059094756	0.735935819	
After sales services	1.381803693	0.057471266	0.71963748	
Conflict resolution	1.21096394	0.044124859	0.6275444	
Occupational safety management	1.487006086	0.06542623	0.776216158	
Social acceptability	1.340629142	0.053482104	0.697055623	

TABLE 6.7 Decision Matrix of Environmental Parameter for Sustainable Supplier Selection Criteria in Mining.

	Quartzite	Lime stone	Coal	Iron	Copper	Gold	Bauxite
Environmental manage	4.2	3.1	3.4	3.2	2.5	2	3
Resource consumption	3.1	3.7	4.1	3.5	3.5	3.4	3.6
Ecological design	3.2	3.8	4.1	4.1	4	4.1	3.8
Recycling	3.9	3.7	3.5	3.7	3.7	3.6	3.5
Ecological impact control	3.45	3.9	4.4	3.8	3.9	3.9	3.6
Sewage water	3.11	4.1	3.9	4	3.7	3.5	4.2
Energy consumption	4.5	3.7	3.5	3.9	4.4	3.8	3.9
Reuse	3.7	3.9	3.6	3.6	4.1	3.8	4
Air emissions	3.1	4	3.6	3.7	3.2	3.7	3.2
Environmental code of conduct	3.9	4	3.5	4.1	3.5	3.7	4.6
Green image	3	3.9	3.4	3.6	2.9	3.3	3.3
Pollution control	4	3.9	4	3.5	3.3	3	3.2
Greenhouse emission	2.5	3.3	3.1	3.2	2.6	4	4.2
Green transportation	3.7	3.6	4.2	3.9	4	3.5	3.3
Hazardous material management	2.8	3.2	2.7	3.3	3.1	3.2	2.6
Noise	4.1	4.2	3.6	3.6	4.2	3.9	4
Max/min	4.5	4.6	4.4	3.2	2.6	3	2.6

TABLE 6.8 Decision Matrix is Normalized for Environmental Parameter.

	Quartzite	Lime stone	Coal	Iron	Copper	Gold	Bauxite
Environmental manage	0.933333333	0.673913043	0.772727273	1	0.961538462	0.666666667	1.153846154
Resource consumption	0.688888889	0.804347826	0.931818182	1.09375	1.346153846	1.133333333	1.384615385
Ecological design	0.711111111	0.826086957	0.931818182	1.28125	1.538461538	1.366666667	1.461538462
Recycling	0.866666667	0.804347826	0.795454545	1.15625	1.423076923	1.2	1.346153846
Ecological impact control	0.766666667	0.847826087	1	1.1875	1.5	1.3	1.384615385
Sewage water	0.691111111	0.891304348	0.886363636	1.25	1.423076923	1.166666667	1.615384615

TABLE 6.8 *(Continued)*

	Quartzite	Lime stone	Coal	Iron	Copper	Gold	Bauxite
Energy consumption	1	0.804 34782 6	0.79 545 454 5	1.21 875	1.69230 7692	1.2666 66667	1.5
Reuse	0.82222 2222	0.847 82608 7	0.81 818 181 8	1.12 5	1.57692 3077	1.2666 66667	1.5384 61538
Air emissions	0.68888 8889	0.869 56521 7	0.81 818 181 8	1.15 625	1.23076 9231	1.2333 33333	1.2307 69231
Environmental code of conduct	0.86666 6667	0.869 56521 7	0.79 545 454 5	1.28 125	1.34615 3846	1.2333 33333	1.7692 30769
Green image	0.66666 6667	0.847 82608 7	0.77 272 727 3	1.12 5	1.11538 4615	1.1	1.2692 30769

TABLE 6.8 *(Continued)*

	Quartzite	Lime stone	Coal	Iron	Copper	Gold	Bauxite
Pollution control	0.888888889	0.847826087	0.909090909	1.09375	1.269230769	1	1.230769231
Greenhouse emission	0.555555556	0.717391304	0.704545455	1	1	1.333333333	1.615384615
Green transportation	0.822222222	0.782608696	0.954545455	1.21875	1.538461538	1.166666667	1.269230769
Hazardous material management	0.622222222	0.695652174	0.613636364	1.03125	1.192307692	1.066666667	1
Noise	0.911111111	0.913043478	0.818181818	1.125	1.615384615	1.3	1.538461538

TABLE 6.9 Weighted Environmental Parameter.

	Quartzite	Lime stone	Coal	Iron	Copper	Gold	Bauxite
Environmental manage	0.23333	0.1684	0.19318		0.24038	0.16666	0.28846
	3333	78261	1818	0.25	4615	6667	1538
Resource consumption	0.17222	0.2010	0.23295	0.273	0.33653	0.28333	0.34615
	2222	86957	4545	4375	8462	3333	3846
Ecological design	0.17777	0.2065	0.23295	0.320	0.38461	0.34166	0.36538
	7778	21739	4545	3125	5385	6667	4615
Recycling	0.21666	0.2010	0.19886	0.289	0.35576		0.33653
	6667	86957	3636	0625	9231	0.3	8462
Ecological impact control	0.19166	0.2119		0.296			0.34615
	6667	56522	0.25	875	0.375	0.325	3846
Sewage water	0.17277	0.2228	0.22159	0.312	0.35576	0.29166	0.40384
	7778	26087	0909	5	9231	6667	6154
Energy consumption		0.2010	0.19886	0.304	0.42307	0.31666	
	0.25	86957	3636	6875	6923	6667	0.375
Reuse	0.20555	0.2119	0.20454	0.281	0.39423	0.31666	0.38461
	5556	56522	5455	25	0769	6667	5385
Air emissions	0.17222	0.2173	0.20454	0.289	0.30769	0.30833	0.30769
	2222	91304	5455	0625	2308	3333	2308
Environmental code of conduct	0.21666	0.2173	0.19886	0.320	0.33653	0.30833	0.44230
	6667	91304	3636	3125	8462	3333	7692

TABLE 6.9 *(Continued)*

	Quartzite	Lime stone	Coal	Iron	Copper	Gold	Bauxite
Green image	0.16666	0.2119	0.19318	0.281	0.27884		0.31730
	6667	56522	1818	25	6154	0.275	7692
Pollution control	0.22222	0.2119	0.22727	0.273	0.31730		0.30769
	2222	56522	2727	4375	7692	0.25	2308
Greenhouse emission	0.13888	0.1793	0.17613			0.33333	0.40384
	8889	47826	6364	0.25	0.25	3333	6154
Green transportation	0.20555	0.1956	0.23863	0.304	0.38461	0.29166	0.31730
	5556	52174	6364	6875	5385	6667	7692
Hazardous material management	0.15555	0.1739	0.15340	0.257	0.29807	0.26666	
	5556	13043	9091	8125	6923	6667	0.25
Noise	0.22777	0.2282	0.20454	0.281	0.40384		0.38461
	7778	6087	5455	25	6154	0.325	5385

TABLE 6.10 Relative Performances of Alternatives Are Calculated According to the WSM Method.

Q1	Q2	Q	Rank	
1.540506233	0.068441192	0.804473712	2nd	Environmental manage
1.845726865	0.09237426	0.969050563		Resource consumption
2.029233229	0.107094449	1.068163839	3rd	Ecological design
1.897987452	0.09738944	0.997688446		Recycling
1.996652035	0.106199556	1.051425795		Ecological impact control
1.980976825	0.102809146	1.041892986	6th	Sewage water
2.069381683	0.111223542	1.090302612	4th	Energy consumption
1.998820352	0.104739253	1.051779803		Reuse
1.80693943	0.089658543	0.948298986		Air emissions
2.040413595	0.108326772	1.074370183	Ist	Environmental code of conduct
1.724208853	0.082667169	0.903438011		Green image
1.809888971	0.091937826	0.950913399		Pollution control
1.731552566	0.07794654	0.904749553		Greenhouse emission
1.938121337	0.101006611	1.019563974		Green transportation
1.55543378	0.067905111	0.811669445		Hazardous material management
2.05529564	0.110850533	1.083073087	5th	Noise

TABLE 6.11 Social Parameter for Sustainable Supply Chain Management of Mining Industry.

	Quartzite	Lime stone	Coal	Iron	Cop per	Gold	Bauxite
Stakeholder engagement	4.2	3.1	3.2	2.8	2.6	2.5	3.1
Staff training	3.1	3.7	4	2.7	3.5	2.7	4.1
Commitment in social	3.2	3.8	4	4.1	4	4	3.2
Commitment to health and safety management	3.9	4.6	3.5	3.7	3.7	3.6	3.5
Stake holder relation	3.45	3.9	4.4	3.8	3.9	3.9	3.4
Code of social conduct	3.11	4.1	3.6	4	3.7	3.5	4.1
Donations for sustainable products	4.6	3.7	3.7	3.9	4.4	2.6	3.6
Rights of interested parties	3.7	3.9	3.6	3.6	4.1	2.6	4
Annual number of accidents	3.1	4	3.6	3.7	3.2	3.7	3.2
Safety practices	3.9	4	3.2	4.1	3.5	3.7	2.9
Rights of stakeholders	3	3.9	3.4	3.6	2.9	3.3	3.3
Human rights	4	3.9	4	3.5	3.3	3	3.2
Health and safety of employees	2.5	3.3	3.1	3.2	2.6	3.7	4.2
Respect for the policy	3.7	3.6	4	3.9	3.7	3.5	3.3
Information disclosure	2.8	3.2	3.2	3.3	3.1	3.2	2.1
Enterprise resource planning	4.1	4.2	3.1	3.6	4.2	4	4
Max/min	4.6	4.6	4.4	2.8	2.6	2.5	2.1

TABLE 6.12 Normalized Value of Social Parameter.

	Quartzite	Lime stone	Coal	Iron	Copper	Gold	Bauxite
Stakeholder	0.913043	0.67391					1.47619
engagement	478	3043	1	0.875	1	1	0476
Staff training	0.673913	0.80434	0.9642	0.8437		1.	1.95238
	043	7826	85714	5	1.346153846	08	0952
Commitment in social	0.695652	0.82608	1.4642	1.2812		1.	1.52380
	174	6957	85714	5	1.538461538	6	9524
Commitment to health and	0.847826		1.3214	1.1562		1.	1.66666
safety management	087	1	28571	5	1.423076923	44	6667
Stake holder relation		0.84782	1.3571			1.	1.61904
	0.75	6087	42857	1.1875	1.5	56	7619
Code of social conduct	0.676086	0.89130	1.4285			1.	1.95238
	957	4348	71429	1.25	1.423076923	4	0952
Donations for sustainable		0.80434	1.3928	1.2187		1.	1.71428
products	1	7826	57143	5	1.692307692	04	5714
Rights of interested parties	0.804347	0.84782	1.2857	1.125	1.576923077	1.	1.90476
	826	6087	14286			04	1905
Annual number of accidents	0.673913	0.86956	1.3214	1.1562	1.230769231	1.	1.52380
	043	5217	28571	5		48	9524
Safety practices	0.847826	0.86956	1.4642	1.2812	1.346153846	1.	1.38095
	087	5217	85714	5		48	2381

TABLE 6.12 *(Continued)*

	Quartzite	Lime stone	Coal	Iron	Copper	Gold	Bauxite
Rights of stakeholders	0.652173	0.84782	1.2857			1.	1.57142
	913	6087	14286	1.125	1.115384615	32	8571
Human rights	0.869565	0.84782		1.0937		1.	1.52380
	217	6087	1.25	5	1.269230769	2	9524
Health and safety of	0.543478	0.71739	1.1428			1.	
employees	261	1304	57143	1	1	48	2
Respect for the policy	0.804347	0.78260	1.3928	1.2187		1.	1.57142
	826	8696	57143	5	1.423076923	4	8571
Information disclosure	0.608695	0.69565	1.1785	1.0312		1.	
	652	2174	71429	5	1.192307692	28	1
Enterprise resource planning	0.891304	0.91304	1.2857			1.	1.90476
	348	3478	14286	1.125	1.615384615	6	1905

TABLE 6.13 Weighted Social Parameter.

	Quartzite	Lime stone	Coal	Iron	Copper	Gold	Bauxite
Stakeholder engagement	0.22826087	0.168478261	0.25	0.21875	0.25	0.25	0.369047619
Staff training	0.168478261	0.20108695 7	0.241071429	0.2109375	0.336538462	0.27	0.488095238
Commitment in social	0.173913043	0.206521739	0.366071429	0.3203125	0.384615385	0.4	0.380952381
Commitment to health and safety management	0.211956522	0.25	0.330357143	0.2890625	0.355769231	0.36	0.416666667
Stake holder relation	0.1875	0.211956522	0.339285714	0.296875	0.375	0.39	0.404761905
Code of social conduct	0.169021739	0.222826087	0.357142857	0.3125	0.355769231	0.35	0.488095238
Donations for sustainable products	0.25	0.201086957	0.348214286	0.3046875	0.423076923	0.26	0.428571429

TABLE 6.13 *(Continued)*

	Quartzite	Lime stone	Coal	Iron	Copper	Gold	Bauxite
Rights of interested parties	0.201	0.211					
	08695	9565	0.32142	0.281	0.3942307		0.4761
	7	22	8571	25	69	0.26	90476
Annual number of accidents	0.168	0.217	0.33035	0.289	0.3076923	0.37	0.3809
	47826	3913	7143	0625	08		52381
	1	04					0.3452
							38095
Safety practices	0.211	0.217					
	95652	3913	0.36607	0.320	0.3365384		
	2	04	1429	3125	62	0.37	
Rights of stakeholders	0.163	0.211	0.32142	0.281	0.2788461		0.3928
	04347	9565	8571	25	54	0.33	57143
Human rights	8	22					
	0.217	0.211					
	39130	9565		0.273	0.3173076		0.3809
	4	22	0.3125	4375	92	0.3	52381
Health and safety of employees	0.135	0.179					
	86956	3478	0.28571				
	5	26	4286	0.25	0.25	0.37	0.5

TABLE 6.13 *(Continued)*

	Quartzite	Lime stone	Coal	Iron	Copper	Gold	Bauxite
Respect for the policy	0.201	0.195					
	08695	6521	0.34821	0.304	0.3557692		0.3928
	7	74	4286	6875	31	0.35	57143
Information disclosure	0.152	0.173					
	17391	9130	0.29464	0.257	0.2980769		
	3	43	2857	8125	23	0.32	0.25
Enterprise resource planning	0.222	0.228					
	82608	2608	0.32142	0.281	0.4038461		0.4761
	7	7	8571	25	54	0.4	90476

TABLE 6.14 Ranking of Social Criteria.

Q1	Q2	Q	Rank	
1.734536749	0.083455787	0.908996268		Stakeholder engagement
1.916207846	0.093493752	1.004850799		Staff training
2.232386477	0.125342614	1.178864545	1st	Commitment in social
2.213812062	0.128190391	1.171001227	2nd	Commitment to health and safety management
2.205379141	0.124071117	1.164725129	3rd	Stake holder relation
2.255355152	0.126425709	1.19089043	6th	Code of social conduct
2.215637094	0.125924507	1.1707808		Donations for sustainable products
2.146143295	0.117105709	1.131624502	8th	Rights of interested parties
2.063933897	0.11097819	1.087456043		Annual number of accidents
2.167508311	0.123451381	1.145479846	4th	Safety practices
1.979381868	0.103088476	1.041235172		Rights of stakeholders
2.013545399	0.109311889	1.061428644		Human rights
1.970931677	0.094721922	1.032826799		Health and safety of employees
2.14826729	0.119539166	1.133903228	7th	Respect for the policy
1.746619237	0.083209449	0.914914343		Information disclosure
2.333802158	0.137137993	1.235470076	5th	Enterprise resource planning

TABLE 6.15 Decision Matrix of Economic Parameter in Mining Sector.

	Quartzite	Lime stone	Coal	Iron	Copper	Gold	Bauxite
Quality	4.2	3.5	3.2	2.5	2.5	2.5	3.5
Flexibility	3.6	3.7	3.2	2.7	3.5	2.7	4
Price	3.2	3.3	4.1	4.1	4	4	3.2
Delivery term	3.5	3.1	3.5	3.7	3.7	3.6	3.5
Relationship	3.1	3.9	4.4	3.8	3.9	3.9	3.4
Cost	2.8	4.6	3.6	4	3.7	3.5	4.1
Technical capacity	4.3	3.7	3.7	3.9	4	2.6	3.6
Logistical cost	3.1	4.3	3.6	3.6	3.6	2.6	4
Reverse logistic	3.1	3.6	3.6	3.7	3.2	3.7	3.2
Rejection rate	3.4	3.6	3.2	4.1	3.5	3.7	2.9
Professionalism	2.8	3.9	3.4	3.6	2.6	3.3	3.3
Product development	3.9	3.4	4	3.5	3.3	3	3.2
Process improvement	2.3	2.9	3.1	3.2	2.6	3.7	4
JIT	2.9	3.6	4	3.9	3.7	3.5	3.3
Commitment	2.8	3	3.2	3.3	3.1	3.2	3.1
Integrity	4	4	3.1	3.6	3.2	4	4
Max/min	4.3	4.3	4.4	2.5	2.5	2.5	3.1

TABLE 6.16 Normalized Economic Parameter.

	Quartzite	Lime stone	Coal	Iron	Copper	Gold	Bauxite
Quality	0.976744186	0.813953488	0.727272727	1	0.961538462	1	1.129032258
Flexibility	0.837209302	0.860465116	0.727272727	1.08	1.346153846	1.08	1.290322581
Price	0.744186047	0.76744186	0.931818182	1.64	1.538461538	1.6	1.032258065
Delivery term	0.813953488	0.720930233	0.795454545	1.48	1.423076923	1.44	1.129032258
Relationship	0.720930233	0.906976744	1	1.52	1.5	1.56	1.096774194
Cost	0.651162791	1.069767442	0.818181818	1.6	1.423076923	1.4	1.322580645
Technical capacity	1	0.860465116	0.840909091	1.56	1.538461538	1.04	1.161290323

TABLE 6.16 *(Continued)*

	Quartzite	Lime stone	Coal	Iron	Copper	Gold	Bauxite
Logistical cost	0.720930233	1	0.818181818	1.44	1.384615385	1.04	1.290322581
Reverse logistic	0.720930233	0.837209302	0.818181818	1.48	1.230769231	1.48	1.032258065
Rejection rate	0.790697674	0.837209302	0.727272727	1.64	1.346153846	1.48	0.935483871
Professionalism	0.651162791	0.906976744	0.772727273	1.44	1	1.32	1.064516129
Product development	0.906976744	0.790697674	0.909090909	1.4	1.269230769	1.2	1.032258065
Process improvement	0.534883721	0.674418605	0.704545455	1.28	1	1.48	1.290322581
JIT	0.674418605	0.837209302	0.909090909	1.56	1.423076923	1.4	1.064516129

TABLE 6.16 *(Continued)*

	Quartzite	Lime stone	Coal	Iron	Copper	Gold	Bauxite
Commitment	0.651162791	0.697674419	0.727272727	1.32	1.192307692	1.28	1
Integrity	0.930232558	0.930232558	0.704545455	1.44	1.230769231	1.6	1.290322581

TABLE 6.17 Weighted Economic Parameter.

	Quartzite	Lime stone	Coal	Iron	Copper	Gold	Bauxite
Quality	0.244						
	18604	0.2034			0.240384		0.28225806
	7	88372	0.181818182	0.25	615	0.25	5
Flexibility	0.209						
	30232	0.2151			0.336538		0.32258064
	6	16279	0.181818182	0.27	462	0.27	5
Price	0.186						
	04651	0.1918			0.384615		0.25806451
	2	60465	0.232954545	0.41	385	0.4	6
Delivery term	0.203						
	48837	0.1802			0.355769		0.28225806
	2	32558	0.198863636	0.37	231	0.36	5
Relations-hip	0.180						
	23255	0.2267					0.27419354
	8	44186	0.25	0.38	0.375	0.39	8
Cost	0.162						
	79069	0.2674			0.355769		0.33064516
	8	4186	0.204545455	0.4	231	0.35	1
Technical Capacity		0.2151			0.384615		0.29032258
	0.25	16279	0.210227273	0.39	385	0.26	1

TABLE 6.17 *(Continued)*

	Quartzite	Lime stone	Coal	Iron	Copper	Gold	Bauxite
Logistical	0.180	0.25	0.204545455	0.36	0.346153	0.26	0.32258064
Cost	23255				846		5
	8						
Reverse logistic	0.180						
	23255	0.2093			0.307692		0.25806451
	8	02326	0.204545455	0.37	308	0.37	6
Rejection rate	0.197						
	67441	0.2093			0.336538		0.23387096
	9	02326	0.181818182	0.41	462	0.37	8
Professionalism	0.162						
	79069	0.2267					0.26612903
	8	44186	0.193181818	0.36	0.25	0.33	2
Product development	0.226						
	74418	0.1976			0.317307		0.25806451
	6	74419	0.227272727	0.35	692	0.3	6
Process improvement	0.133	0.1686					0.32258064
	72093	04651	0.176136364	0.32	0.25	0.37	5
JIT	0.168						
	60465	0.2093			0.355769		0.26612903
	1	02326	0.227272727	0.39	231	0.35	2

TABLE 6.17 *(Continued)*

	Quartzite	Lime stone	Coal	Iron	Copper	Gold	Bauxite
Commitment	0.162						
	79069	0.1744			0.298076		
	8	18605	0.181818182	0.33	923	0.32	0.25
Integrity	0.232	0.2325			0.307692		0.32258064
	55814	5814	0.176136364	0.36	308	0.4	5

TABLE 6.18 Ranking of Economic Criteria.

Q1	Q2	Q	Rank	
1.65213528	0.078674266	0.865404773		Quality
1.805355893	0.089716288	0.94753609		Flexibility
2.063541423	0.107862194	1.085701809	3rd	Price
1.950611862	0.09938293	1.024997396		Delivery term
2.076170293	0.111702149	1.093936221	1st	Relationship
2.071192405	0.110046656	1.09061953	2nd	Cost
2.000281517	0.106368232	1.053324875		Technical capacity
1.923512504	0.099068731	1.011290618		Logistical cost
1.899837162	0.095699877	0.997768519		Reverse logistic
1.939204355	0.097350819	1.018277587		Rejection ratse
1.788845734	0.086645333	0.937745534		Professionalism
1.87706354	0.096740435	0.986901988		Product development
1.74104259	0.078471508	0.909757049		Process improvement
1.967077967	0.100901136	1.033989551		JIT
1.717104407	0.079835858	0.898470133		Commitment
2.031525596	0.1080208	1.069773198	4th	Integrity

It is found that, supplier selection criteria for all mining sector technical position is ranked 1st, and the impression and financial position are ranked 2nd and 3rd, respectively.

Similarly by considering sustainable supply chain management of Indian mining industries, under environmental parameter, environmental code of conduct, environmental manage, and ecological design are highest ranked, under economic parameter cost, relation and price are the most important criteria ranked highest, whereas under social criteria, commitment in social, commitment to health and safety management and stakeholder relation ranked highest. Supply chain administration can permit resource, energy, and transport expenses to be systematically compiled. These charges can always reduce when agencies diagram their personal tactics (for example, logistics) extra successfully and/or help their suppliers in setting up or enhancing administration process. Additionally, extra environment friendly procedures and structures minimize the demand for substances and consequently manufacturing costs. Responsible occupational fitness and security practices can additionally lead to price savings via elevated productiveness and fewer accidents. Anchoring sustainability concepts in product improvement can fortify ability for innovation. Innovative options for manufacturing procedures and transport can additionally be discovered by using working with suppliers on these issues.

6.4 CONCLUSION

For any supply chain network, one of the most sizeable choices is deciding on fabulous suppliers. Suppliers collect and supply uncooked substances or different assets to execute provide chain functions. Traditional dealer resolution is primarily based on price, quality, and lead time, which may additionally no longer healthy into the contemporary commercial enterprise environment. In this study, WASPAS approach used to be proposed for sustainable dealer selection. The sustainable provider choice procedure can enhance relationships with every stage of customer, due to the fact environmental administration standards make sure eco-friendly merchandize for sustainably mindful consumers.

KEYWORDS

- **challenges**
- **government**
- **mining**
- **supplier**
- **operations**
- **alternative**

REFERENCES

Alegoz, M.; Yapicioglu, H. Supplier Selection and Order Allocation Decisions under Quantity Discount and Fast Service Options. *Sustain. Prod. Consump.* **2019,** *18,* 179–189.

Arabsheybani, A.; Paydar, M. M.; Safaei. A. S. An Integrated Fuzzy MOORA Method and FMEA Technique for Sustainable Supplier Selection Considering Quantity Discounts and Supplier's Risk. *J. Clean. Prod.* **2018,** *190,* 577–591.

Çebi, F.; Otay, İ. A Two-stage Fuzzy Approach for Supplier Evaluation and Order Allocation Problem with Quantity Discounts and Lead Time. *Inf. Sci.* **2016,** *339,* 143–157.

Cheraghalipour, A.; Farsad, S. A Bi-objective Sustainable Supplier Selection and Order Allocation Considering Quantity Discounts under Disruption Risks: A Case Study in Plastic Industry. *Comput. Ind. Eng.* **2018,** *118,* 237–250.

Ghadimi, P.; Toosi, F. G.; Heavey, C. A Multi-agent Systems Approach for Sustainable Supplier Selection and Order Allocation in a Partnership Supply Chain. *Eur. J. Oper. Res.* **2018,** *269* (1), 286–301.

Hamdan, S.; Cheaitou, A. Dynamic Green Supplier Selection and Order Allocation with Quantity Discounts and Varying Supplier Availability. *Comput. Ind. Eng.* **2017a,** *110,* 573–589.

Hamdan, S.; Cheaitou, A. Supplier Selection and Order Allocation with Green Criteria: An MCDM and Multi-objective Optimization Approach. *Comput. Oper. Res.* **2017b,** *81,* 282–304.

Moheb-Moheb-Alizadeh, H.; Handfield, R. Sustainable Supplier Selection and Order Allocation: A Novel Multi-objective Programming Model with a Hybrid Solution Approach. *Comput. Ind. Eng.* **2019,** *129,* 192–209.

Rezaei, A.; Rahiminezhad Galankashi, M.; Mansoorzadeh, S.; Mokhatab Rafiei, F. Supplier Selection and Order Allocation with Lean Manufacturing Criteria: An Integrated MCDM and Bi-objective Modelling Approach. *Eng. Manag. J.* **2020,** *32* (4), 253–271.

CHAPTER 7

Assessment of Risk in the Mining Sector of India

ABSTRACT

Mining activities cause a negative environmental effect and social war; but in addition, mining activities provide monetary profit for network and supply the minerals needed for low-carbon energy production. Mining in certain situations has drawn unique attention because in addition to other industrial activities it tends to have a strong terrible effect on the environment and society with regard to minerals and mineral production. However, the risks associated with mining sector and to visitors are also very high. Risk of hazards, accidents are maximum in the mining sector. Thus, an effort is made to study and prioritize the risk factors associated with the mining industry.

7.1 INTRODUCTION

Growing pressure from stakeholders, government authorities, and non-governmental companies are forcing the mining industry to take initiative for the green supply chain management (GSCM). The successful implementation of GSCM is vital for industries to increase the economic–environmental performance and make business sustainability. Gandhi et al. (2015) have evaluated the crucial aspect related to the successful implementations of GSCM. Operations in the mining and quarrying industries have sizable unfavorable environmental results. Mining businesses have started to implement the GSCM that embraces the green information technology

Sustainable Supply Chain Management in the Mining Industry.

Suchismita Satapathy and Hullash Chauhan (Authors)

systems (GITS) to assist in generating monetary benefits while looking for minimum environmental damage. Bai et al. (2017) have addressed this problem by enforcing a framework of GSCM practices and followed a new decision-assisting approach that integrates grey numbers with DEMATEL and the NK model to evaluate and develop an implementation model. The construction industries are the largest waste generators in the world, and to address this problems, the idea of environmental sustainability inside the production sector can be used in unique ways. The GSCM concept has been correctly implemented in other organizations; however, the construction industries have their peculiarity, which calls for a specific method to implement those concepts, which take into account the variations between the sectors. Benachio et al. (2019) have diagnosed the model of GSCM may be applied within the building industries. Meyer et al. (2021) have analyzed how mining can afford information about the effect of the coronavirus pandemic on supply chains (SC) with an emphasis on the effect of the pandemic on SC constructs in relation to risk, resilience, and sustainability. Sustainable supply chain management (SSCM) makes a focus of reducing the terrible effect of operations inside the supply chain in addition to minimizing the strength and material intake. A successful implementation of SSCM is vital for the mining industry to achieve growth in profitability and ecological performance. However, the adoption price of SSCM with the aid of Indian companies, particularly inside the mining quarter is not encouraging (Muduli et al., 2017), developing, comparing, evaluating, and selecting essential GSCM practices. The goal of adoption of the price is to implementat the GSCM. Kusi-Sarpong et al. (2016) have proposed inclusive suggestion of the main practices and sub-practices (framework) of the GSCM that determines the relative relationships and effects within this (GSCM) framework and identifies the perceived influences of the GSCM framework on sustainable industry performance (monetary, ecological, and social-triple soil) which might be relevant to the mining industry in the emerging economies of Ghana. However, sustainable control practices are required to gain long-term benefits. Kusi-Sarpong et al. (2014) have introduced a sizable framework for SSCM which the mining industries in Ghana are adopting. Particularly, in the mining and extractive industries, there are serious and hazardous environmental issues.. A green way to deal with these harmful environmental issues is through organizational exercise that includes the wider supply chain. The green supply chain practices and their role within the method and operation of the mining enterprise have no longer been absolutely addressed (Kusi-Sarpong et al., 2015). Mining occupation causes negative effect on the environmental and social conflicts; However, it also presents the financial advantages for groups and guarantees the minerals needed for low-carbon technologies. Ivic et al. (2021) have analyzed, evaluated, and critically investigated the sustainability report of 10 European mining industries for the period 2016–2018 in an effort to resolve the drivers for adopting the sustainability practices and their contributions to the sustainable development goals (SDG).

The mining industry is well-known throughout the world for its highly volatile and risky operating environment. Technological advancements in ore extraction techniques for the expansion of production stages have created a new challenge for security in this enterprise. To date, research in the field of safety has discovered that the majority of incidents in risky industries are caused by human errors, the supervision of which could significantly improve the safety levels in working environments. The mining enterprise operates under the well-identified truth that it has the hardest working environment, where the protection and fitness of the worker are always a high priority. Mining safety has constantly drawn the attention of researchers working in the area of fitness and safety. The metal and mining enterprise of India has recorded a strong expansion in the recent past, and it is anticipated that India would become the second-largest steel producer by 2015. Preconditions for dangerous activities are in addition categorized into environmental (physical and technical surroundings), operator's circumstance, and personnel factors. The mining enterprise is known for its dynamic and difficult environmental situations. Troubles involved in illumination, air flow, etc., were a hurdle in retaining protection at the worksite. The mining lifecycle includes exploration, mine improvement, mine operations, closure and land reclamation. Mining is a multidisciplinary industry employing a wide range of trades and trades. To make certain precision in scientific and epidemiological paintings, it's far vital to enquire approximately the info of responsibilities, as the time-period miner' is distinctly nonspecific. Mining is historically categorized as metalliferous or coal mining and as surface or underground mining. Metalliferous mining also can be labeled consistent with the commodity being mined. A few degree of minerals processing is generally undertaken at mine sites. For metalliferous mining, many of the occupational fitness hazards are related to these metallurgical approaches, and for this reason, I'm able to include comments on metallurgical hazards. Hazard management and sustainability have grown to be imperative to the way organizations function and occupational fitness needs to be included into these systems. Multinationals are developing global company occupational safety and hygiene standards, which require core components well suited with local legislative requirements. Powerful health surveillance and organic monitoring require sophisticated information frameworks which are best advanced currently.

Mining disasters, firedamp and coal-dust explosions, the inrush of water, fire, and unexpected outbursts (of CO_2 or CH4) are collective risks which have been known in the mining industry from its very inception. They have got always stuck with the creativity, as is proven via the wealth of fiction writing on the situation, in particular from the 19th century. These large-scale accidents very soon brought the systematic creation of preventive measures on a collective and regular basis at a time when different branches of enterprise were now not receiving the same interest. The principle occupational sicknesses affecting miners are of the cumulative type: arthrosis, deafness, pneumoconiosis, and so forth. Current cases of arthrosis updated are few in wide variety and mild in intensity. Pneumoconiosis is a specifically severe problem throughout the world's mining industries, the mining industry has a recognition for being a volatile business, with health risks which can be varied and regularly quite extreme, and it is very crucial for miners to protect themselves.

Though, mining is not up-to-date unsafe. With the advent of strict protection law, up-to-date advances in safety gadgets, the industry has seen its death rate drop over the years.

Although the goal of zero risk has not been performed, it remains the same old that mining businesses retain to strive up. People who work in mines are updated about the growth of health risks that fall under this broad class. Even as musculoskeletal harm can occur during a ride, fall or hard climb, the more severe ones occur slowly through the years. This can be due to ongoing heavy lifting or repetitive lines, says Clark. Inside the mining area, WBV can be prompted either by spending a lot of time sitting on machinery, that is maximum of the time in mining extraction, or by means of status, up-to-date running on jumbo operaup to daters, or by spending a lot of time sitting on equipment, which is the greatest amount of time in mining extraction. Preventing MSDs needs to be updated as a key part of every workplace health and safety program. In secure and healthful workplaces, employers currently discover and assess job-associated MSD hazards and put in area controls to lessen people' exposure to MSD dangers. Regardless of the chemical compounds you figure in close proximity up-to-date, suitable safety wear and precautions must be taken to minimize your frame's exposure to them. Dangers consist of chemical burns, respiration troubles and poisoning, Clark outlines. Over exposure to ultraviolet rays can cause pores and skin cancer. Australia has the best international fee. UV rays can no longer only cause melanomas to change in shape; however, they can harm your eyes if you do not wear protective eyewear. To guard employees from noise, mining organizations compare operating conditions and noise exposure through danger tests. Heading off and decreasing the exposure may be carried out via applying engineering controls on the noise source or alongside the noise route to lessen the exposures, consisting of vibration dampeners or absorptive panels. Everyday renovation of machines is also important to reducing noise. Organization must make sure that proper use of private safety systems among noise-exposed employees, while providing important health and safety training and retaining health surveillance data. Every chemical has a completely unique set of hazards and must be dealt with nicely to ensure certain employee safety. Employers want to conduct danger tests and updated best practices. Airflow is likewise a crucial issue in minimizing exposure, as well as popular home tasks and cleanliness. Thorough education and drills are to be performed concerning the organization's spill reaction plans and chemical hygiene plans.

7.2 MINING AND RISK

Extraction of precious minerals or other geological materials from the earth is referred to as mining, and it often occurs from an ore frame, vein, or seam (Pule, 2011). Mining exposes individuals to potentially hazardous situations and necessitates the use of potentially hazardous tools and materials, whether that is done underground or at ground level (Pule, 2011), with a high prevalence of harm being noted throughout all mining divisions (Ghosh et al., 2004). Anyone who works in a mine may be exposed to a variety of physical, chemical, mechanical, organic, and social dangers (Pule, 2011). For instance, noise may be a major issue due to the increased mechanization of mining and is present everywhere due to a variety of techniques, including boring, drilling, blasting, cutting, handling materials, air movement, crushing, conveying, and ore processing (Donoghue, 2004; Kumar et al., 1998). Without sufficient ear protection, high levels of noise can damage the middle and inner ear canals and impair hearing all at once; for some people, this impairment begins at 25 dBA (Genesove, 2010; Ramchandar and Singh, 2001). Despite the fact that exposure to noise poses a significant danger, accidents and dust are the main causes of mortality and morbidity in mines (McBride, 2004). Rock slides, fires, explosions, mobile system mishaps, falls from high places, trapping, and electrocution are examples of frequent mishaps (Donoghue, 2004). Dust is a common component of mining because the activity involves removing soil and breaking rocks (Pule, 2011). The pebbles and dirt that arise from this might cause physical and physiological harm. If the eyes are not properly protected, dirt can enter them, causing bad eyesight as well as a number of severe skin irritations. Various breathing disorders can result from dirt inhalation. The world recognizes mining as one of the most dangerous industries (ILO, 2010). It is very challenging to look at the safety environment in this area.

The emphasis on saving the environment from stakeholder, with social organizations, authorities, and most people is increasing. As an end-result, constructing a sustainable supply chain control is now part of corporate social obligation and a task for agencies, along with small and medium-sized organizations (Wu et al., 2018). Rebs et al. (2018) have examined the stakeholder risks and dangers on SSCM and this has been addressed by conceptual frameworks, empirical research, and formal model to highlight the tendencies and gaps in qualitative and quantitative studies. Developing awareness of social, monetary, and environmental issues has led corporations to implement sustainable practices. In recent years, corporations were held answerable for their organizational activities which have an effect on the environment, society, and the economy of their own commercial enterprise in addition to their supply chain members. Diverse stakeholders are placing pressure on corporations, and their SC must take sustainable action so that you can achieve the goals of SSCM (Saeed et al., 2017). SSCM has improved appreciably in the fields of mining, commercial enterprises and technology. Sánchez-Flores et al. (2020) have reviewed the existing literature on SSCM in the emerging markets from a worldwide perspective. Mining is critical and an important source of income for the economies of many developing countries. Environmental, social, and economies, all these factors impact on mining. Some factors that impact the mining sector are air and gas emission, waste generations, and health impact, etc.

The health of the mine workers is the most important factor among all problems faced in mining. Nowadays, technological development and the improvement of new strategies lower the degree of coal mining injuries and environmental impact. However, there are not any enough validated research that gifts a complete assessment of protection issues, environmental impact, and the effect of generation in lowering coal mining injures. Therefore, knowing the protection and environmental concerns in coal mining allows to increase the diverse forms of strategies for enhancing protection, equipment layout, work approaches, work schedules, safety packages, and emergency reaction plans, and strategies for diagnosing the capacity risks associated with new technologies and paintings approaches.

The National Institute for Occupational Safety and Health (NIOSH) has distinct health and safety applications targeted on mining and oil and gasoline extraction (OGE), both of which have established disparities in health effects (Yeoman et al., 2016; NIOSH, 2015). NIOSH has validated the prevalence of coal workers' pneumoconiosis (CWP), which has been increasing since 2000, and 1 in 10 long-tenured miners show the evidence of having CWP (Blackley et al., 2018). Miners other than coal people additionally have disparities in fitness consequences, with some research demonstrating that steel and nonmetal miners have excess mortality due to cardiovascular disease, in addition to increased risk of lung cancers, nonmalignant respiration disorder, and chronic breathing signs (Yeoman et al., 2016). Elevated mortality from prostate cancer has been diagnosed among some workers employed in the U.S oil businesses, despite the fact that no upward trend has been observed with extended duration of employment (Wong and Raabe, 2000). Records from NIOSH's country-wide Occupational Mortality Surveillance program has confirmed that U.S. OGE workers have shown increased mortality from coronary heart sickness and several cancer sorts (i.e., larynx, prostate, oral hollow space, pharynx, bladder, kidney, and mesothelioma; NIOSH, 2015).

7.3 METHODOLOGY

A study has been conducted to find the risk factors faced by the Indian mining workers working in mining sector. A standard questionnaire is designed and a survey is conducted to find risk factors faced by workers or occupational risk faced in Indian mining industries. The principal component analysis is conducted in combination with the WASPAS method to prioritize risk factors in mining sector. A team of researchers was sent to different mining sectors, such as coal, bauxite, limestone, and quartz mines of India. A total of 200 questions were sent to different mines in India, and among them, 140 data were collected (Appendix 1). The data are rated by Likert scale (1–5), where 1 = agree, 2 = somewhat agree, 3 = no pinion, 4 = somewhat disagree, 5 = disagree.

7.4 RESULTS AND DISCUSSION

Minitab 17 is used for conducting the factor analysis of the collected 140 data of mining risk. It was found that eight dimensions are selected whose Chrobanch 'alpha value is more than 0.5, rest are discarded. The eight dimensions are explosion, falling structure, fires, dust, noise, centricity hazard, vehicle hazard, etc.

Mines can frequently present much more dangerous hazards than your regular artificial installation. The following is a list of just some of the perils miners can face. This list is by no means complete, but rather collected to serve as a pointer to the most common hazards in a mine.

7.4.1 EXPLOSIONS

The trouble of an implicit explosion generally occurs in underground mines and poses a serious trouble to safety. Because ignitable chemicals and feasts are frequently passing through pipes in an underground mine, it is essential that the ventilation systems are periodically inspected, and warning signs are posted to remind workers of safe practices.

7.4.2 FALLING STRUCTURES

The infrastructure of mines can occasionally result in catastrophe. The roof of a mine can collapse and can cause rocks and debris to become unstable and conceivably strike workers. Occasionally, collapses are listed and totally planned to control the hazard of a cave-in. Proper safety plans will need to be practiced, and mine workers should be trained on what to do in the case of a cave-in or a collapse in the mine.

7.4.3 FIRES

A number of objects and areas in a mine can cause a fire, and traditional fire suppression systems may not always be suitable in certain mining environment. Because of this, it'll be especially important to concentrate on fire hazards and fire safety in the overall mining health and safety program.

7.4.4 ELECTRICAL HAZARDS

Most mines have different pieces of electrical equipments used on a regular basis. Although necessary, these machines can pose a threat of fire, shock, or bow flash. And the cords connected to the electrical equipments also pose a trip and fall hazard to workers. Programs should be put in place and electrical safety training should be offered to miners. Specific training may need to take place for especially dangerous machines.

7.4.5 DUST EXPOSURE

Miners, especially coal miners, care at threat for breathing in dust and debris while working and consequences can be serious. Safety directors and employers will need to assess their plant and give proper respiratory protection.

7.4.6 VEHICLE HAZARDS

Artificial vehicles, such as cement exchanges, cranes, or tractors can frequently be seen at mining spots. These vehicles do not generally offer the stylish visibility for drivers and can put all the workers in the area at threat. Ensure workers can be fluently spotted with hi-vis safety vests and take the time to draw out a traffic plan for the area.

7.4.7 NOISE

The loud equipment used in the confined spaces of a mine can cause endless hearing loss. Machines, such as roof bolters, bulldozers, nonstop mining machines, frontal end payload, and others are responsible for inordinate noise. The correct hearing protection will need to be supplied to keep workers from passing unrecoverable damage.

TABLE 7.1 Factor Analysis.

Factors	F1	F2	F3	F4	F5	F6	F7	F8
Explosion	0.657							
Falling structure		0.765						
Fires			0.511					
Electric hazards				0.509				
Dust exposure					0.728			
Vehicle hazard						0.603		
Noise							0.544	
Poisonous gas/liquid								0.567

The selected factors are ranked by WASPAS method.

WASPAS technique:

The WASPAS technique is a completely unique aggregate of well-known MCDM approaches, that is, its software first calls for improvement of a decision/evaluation matrix, X = [xij]m×n in which xij is the overall performance of ith opportunity with admire to jth criterion, m is the variety of options and n is the variety of criteria. To make the overall performance measures similar and dimensionless, all the factors inside the selection matrix are normalized with the use of the subsequent equations:

Step 1

for beneficial criteria

xij=min i xij/xij

for non-beneficial criteria,

where xij is the normalized value

Step 2

Qi(1)=_j=1^n(XijWj)

in which wj is the weight

(relative significance) of jth criterion.

Qi(2)=_j1^nxijwj. A joint generalized criterion of weighted aggregation of additive and multiplicative methods is then proposed by Saparauskas et al. (2011): In order to have increased rating accuracy and effectiveness of the decision-making process, in WASPAS method, an extra generalized equation for figuring out the overall relative significance of ith opportunity is developed (Zavadskas et al., 2012; Zavadskas et al., 2013a, 2013b) as given below:

Qi=lamba*Qi(1)+(1-lambda)Qi(2)

The feasible options are now ranked primarily based totally at the Q values and the excellent opportunity has the maximum Q cost. In eq 7 (Stević et al., 2017), while the cost is 0, WASPAS method is converted to WPM, and while is 1, it will become WSM method. It has been carried out for fixing MCDM issues for growing rating accuracy, and it has the functionality to attain the very best accuracy of estimation (Bagocius et al.)

TABLE 7.2 Normalization Matrixes.

	Quartz/limestone	**Coal**	**Bauxite**
Explosion	2.875	4.1875	3.833333333
Falling structures	3.8125	4.1875	4.166666667
Fires	4.375	2.9375	3.833333333
Electrical hazards	4.75	3.5	4.055555556
Dust exposure	4.1875	3.1875	3.388888889
Vehicle hazard	4.375	2.375	4.277777778
Poisonous gas/liquid	4.0625	3.0625	3.833333333

According to the results of the WASPAS method, the most important risk factor is the falling structure, followed by dust exposure and electric hazards. As a result, steps must be taken to reduce the risk of mining structures, so that the workers' and visitors' lives are safe and risk-free.

Then correlation analysis is conducted to find the correlation among risk factors in mining sector.

TABLE 7.3 Weighted Aggregated Sum Product Assessment (WASPAS).

	Quartz/limestone	Coal	Bauxite
Explosion	1	0.686567164	0.75
Falling structures	1	0.910447761	0.915
Fires	0.671428571	1	0.766304348
Electrical hazards	0.736842105	1	0.94057377
Dust exposure	0.76119403	1	0.94057377
Vehicle hazard	0.542857143	1	0.555194805
Poisonous gas/liquid	0.753846154	1	0.798913044

Weight = 0.25, lambda = 0.05.

TABLE 7.4 Shows Ranking of Risk Factors.

	Q1	Q2	Q
Explosion	0.609141791	0.847102714	0.728122252(6)
Falling structures	0.70636194	0.955364351	0.830863145(1)
Fires	0.691576087	0.846935383	0.769255735(5)
Electrical hazards	0.735143443	0.912413757	0.8237786(3)
Dust exposure	0.735143443	0.919860689	0.827502066(2)
Vehicle hazard	0.638798701	0.740939482	0.689869092(8)
Poisonous gas/liquid	0.699728261	0.880938436	0.790333348(4)

The correlation of risk factor shows that the risk factor explosion shows highest correlation and falling structure is showing lowest among all risk factors.

7.5 CONCLUSION

Regardless of improvements in safety overall performance, the quantity and severity of mining-associated injuries remain high and unacceptable, indicating that further discount may be accomplished. Protection issues in coal mining had been operational factors, human factors, device factors, environmental factors, and absence of detailed emergency rescue plans. Those threat-based processes frequently use precise key phrases and ideas, consisting of hazard, threat, events, and controls. Threat evaluation involves the assigning of a stage of threat to every of the common health and protection hazards at a place of job, followed via the rating of those dangers. A hazard is any source of reduced ability, damage, or destructive effect on the health of something or someone at work under certain conditions. Danger is typically defined as the danger of an event that could affect targets. It is also evaluated in terms of the likelihood and consequences of the event. Dangers can be inherent or residual. While such situations or conditions do clearly materialize in this sort of manner as to jeopardize the health and protection of people, they may be known as activities.

TABLE 7.5 Shows Correlation of Risk Factor.

Parameter	Correlation of risk factor
Explosion	0.579*
Falling structures	0.076
Fires	0.215
Electrical hazards	0.123
Dust exposure	0.514*
Vehicle hazard	0.125
Noise	0.522*
Poisonous gas/liquid	0.358

As soon as it is identified, offices want to determine the correct controls to prevent or mitigate the threat associated with a risk. Controls may be sports, methods, bodily device or gadgets, or systems. Controls are every so often described as stopping (i.e., reducing the probability that an undesirable event will arise) and mitigating (i.e., decreasing or doing away with the results of an undesirable occasion), and as critical or non-important. Offices trying to learn more approximately these concepts can refer to the sources indexed at the end of this guiding principle.

IoT and AI techniques may be used for tracing of materials and process to reduce danger in mining sector.

KEYWORDS

- **activities**
- **practices**
- **organizations**
- **developing**
- **companies**
- **occupational**

REFERENCES

Bai, C.; Kusi-Sarpong, S.; Sarkis, J. An Implementation Path for Green Information Technology Systems in the Ghanaian Mining Industry. *J. Clean. Prod.* **2017,** *164,* 1105–1123.

Bazi, S.; Benzid, R.; Bazi, Y.; Rahhal, M. M. A. A Fast Firefly Algorithm for Function Optimization: Application to the Control of BLDC Motor. *Sensors* **2021,** *21* (16), 5267.

Benachio, G. L. F.; Freitas, M. C. D.; Tavares, S. F. In *Green Supply Chain Management in the Construction Industry: A Literature Review*, IOP Conference Series: Earth and Environmental Science, Vol. 225, No. 1; IOP Publishing, 2019; p 012011.

Blackley, D. J.; Halldin, C. N.; Laney, A. S. Continued Increase in Prevalence of Coal Workers' Pneumoconiosis in the United States, 1970-2017. *Am. J. Public Health* **2018,** *108,* 1220–1222.

Donoghue, A. M.; Occupational Health Hazards in Mining: An Overview. *Occup. Med.* **2004,** *54,* 283–289.

Gandhi, S.; Mangla, S. K.; Kumar, P.; Kumar, D. Evaluating Factors in Implementation of Successful Green Supply Chain Management Using DEMATEL: A Case Study. *Int. Strateg. Manag. Rev* . **2015,** *3* (1–2), 96–109.

Genesove, L. In *Occupational Health Hazards in Mining & Metallurgical Work*, Paper Presented at the Mining Health and Safety Conference 2010 in Sudbury, Ontario, Canada, April 20–22, 2010.

Ghosh, A. K.; Bhattacherjee, A.; Chau, N. Relationships of Working Conditions and Individual Characteristics with Occupational Injuries: A Case-control Study in Coal Miners. *J. Occup. Health* **2004,** *46,* 470–478.

Ghosh, A. K.; Bhattacherjee, A.; Ray, S. C. An Application of System Dynamics in Mine Safety Studies. *Miner. Resour. Eng.* **1998,** *7,* 131–147.

ILO, 2010. Mining: A Hazardous Work; January 07, 2013, 2010.

Ivic, A.; Saviolidis, N. M.; Johannsdottir, L. Drivers of Sustainability Practices and Contributions to Sustainable Development Evident in Sustainability Reports of European Mining Companies. *Discov. Sustain.* **2021,** *2* (1), 1–20.

Kusi-Sarpong, S.; Bai, C.; Sarkis, J.; Wang, X. Green Supply Chain Practices Evaluation in the Mining Industry Using a Joint Rough Sets and Fuzzy TOPSIS Methodology. *Resour. Policy* **2015,** *46,* 86–100.

Kusi-Sarpong, S.; Sarkis, J.; Wang, X. Assessing Green Supply Chain Practices in the Ghanaian Mining Industry: A Framework and Evaluation. *Int. J. Prod. Econ*. **2016,** *181,* 325–341.

Kusi-Sarpong, S.; Sarkis, J.; Wang, X.; Filho, W. L. Sustainable Supply Chain Management Practices in Ghana's Mining Industry. 2014.

Meyer, A.; Walter, W.; Seuring, S. The Impact of the Coronavirus Crisis on Supply Chains and Their Sustainability: A Text Mining Approach. *Front. Sustain*. **2021,** *2,* 3.

Muduli, K.; Biswal, J. N.; Satapathy, S.; Barve, A.; Tripathy, S. Investigation of Influential Factors of Green Supply Chain Management in Indian Mining Industries: An Empirical Study. *Int. J. Bus. Excell.* **2017,** *12* (3), 351–375.

National Institute for Occupational Safety and Health. National Occupational Mortality Surveillance (NOMS) [Online], 2015. https://www.cdc.gov/niosh/topics/noms/noms2 charts/oil_gas_extraction/ind-nc-48.html.Google Scholar

Nayak, J.; Naik, B.; Dinesh, P.; Vakula, K.; Dash, P. B. Firefly Algorithm in Biomedical and Health Care: Advances, Issues and Challenges. *SN Comput. Sci*. **2020,** *1* (6), 1–36.

Pule, T. Mining Activities and Occupational Health and Safety at Work. *Afr. Newsl. Occup. Health Saf.* **2011,** *21* (1), 4–7.

Ramchandar, K.; Singh, T. N. Dust Generation and Its Control in Coal Mine. *Min. Eng. J.* 2001, *3* (4), 22–26.

Rebs, T.; Brandenburg, M.; Seuring, S.; Stohler, M. Stakeholder Influences and Risks in Sustainable Supply Chain Management: A Comparison of Qualitative and Quantitative Studies. *Bus. Res.* **2018,** *11* (2), 197–237.

Saeed, M. A.; Waseek, I.; Kersten, W. In *Literature Review of Drivers of Sustainable Supply Chain Management*, Proceedings of the Hamburg International Conference of Logistics (HICL), 2017; pp 159–184; epubli.

Sánchez-Flores, R. B.; Cruz-Sotelo, S. E.; Ojeda-Benitez, S.; Ramírez-Barreto, M. Sustainable Supply Chain Management—A Literature Review on Emerging Economies. *Sustainability* **2020,** *12* (17), 6972.

Wong, O.; Raabe, G. K. A Critical Review of Cancer Epidemiology in the Petroleum Industry, with a Meta-analysis of a Combined Database of More than 350,000 Workers. *Regul. Toxicol. Pharmacol.* **2000,** *32,* 78–98.

Wu, Z.; Zhai, S.; Hong, J.; Zhang, Y.; Shi, K. Building Sustainable Supply Chains for Organizations based on QFD: A Case Study. *Int. J. Environ. Res. Public Health* **2018,** *15* (12), 2834.

Yeoman, K. M.; Halldin, C. N.; Wood, J.; Storey, E.; Johns, D.; Laney, A. S. Current Knowledge of US Metal and Nonmetal Miner Health: Current and Potential Data Sources for Analysis of Miner Health Status. *Arch. Environ. Occup. Health* **2016,** *71,* 119–126.

Conclusion

Despite enhancements in overall performance of safety, the frequency and severity of mining-associated injuries remain high and unacceptable, indicating that further reductions may be achieved. Operational factors, human factors, device factors, environmental factors, and the absence of detailed emergency rescue plans have all been identified as coal mining safety concerns. Those threat-based processes frequently use precise key phrases and ideas, consisting of hazard, threat, events, and controls. Threat evaluation involves assigning of a stage of threat to common health and safety hazards at a workplace, followed by assessing those risks. A hazard is any source of capacity damage, damage, or destructive health effects on something or someone under certain situations at work. Danger is usually taken into consideration to be the threat of something that can have an impact on goals. It is also measured in terms of event likelihood and effects. Dangers may be inherent or residual. While such situations or conditions do clearly materialize in this manner so as to jeopardize the health and protection of people, they may be known as activities. Once danger is identified, workplaces should determine the correct controls to prevent or mitigate the threat associated with a risk. Controls may be sports, methods, bodily device or gadgets, or systems. Controls are often described as stopping (i.e., reducing the probability that an undesirable event will arise) and mitigating (i.e., decreasing or doing away with the results of an undesirable occasion), and as critical or non-important. Companies trying to learn more about these concepts can refer to the sources indexed at the end of the guiding principles.

Internet of things and artificial intelligence techniques may be used for tracing of materials and process to reduce dangers in the mining sector and improve workplace safety during supply chain of raw materials in a sustainable manner.

Index

N

O

P

T

U

V

W